# ALFRED RUSSEL WALLACE

**John van Wyhe** is a historian of science, Senior Lecturer in the Department of Biological Sciences and a Fellow of Tembusu College at the National University of Singapore. He is the founder and Director of *Darwin Online* and *Wallace Online*, Professorial Fellow of Charles Darwin University, Fellow of the Linnean Society of London and a Scientific Associate of the Natural History Museum (London). He lectures and broadcasts on Darwin, Wallace, and the history of science around the world.

**Kees Rookmaaker** is a biologist specialising in the history of zoology. He has worked for the past eight years on Darwin and Wallace, including bibliography and transcriptions of notebooks and letters. He has also edited detailed surveys of all letters received by the Museum of Zoology, University in Cambridge during the 19th century. He is the author of over 200 papers and several books. He received the Founder's Medal of the Society for the History of Natural History.

# ALFRED RUSSEL WALLACE:

# LETTERS FROM
# THE MALAY ARCHIPELAGO

Edited by John van Wyhe and Kees Rookmaaker

OXFORD
UNIVERSITY PRESS

# OXFORD

UNIVERSITY PRESS

Great Clarendon Street, Oxford, OX2 6DP,
United Kingdom

Oxford University Press is a department of the University of Oxford.
It furthers the University's objective of excellence in research, scholarship,
and education by publishing worldwide. Oxford is a registered trade mark of
Oxford University Press in the UK and in certain other countries

First published 2013
First published in paperback 2015

British Library Cataloguing in Publication Data
Data available

Library of Congress Cataloging in Publication Data
Data available

ISBN 978-0-19-968399-4 (Hbk.)
ISBN 978-0-19-968400-7 (Pbk.)

Printed in Great Britain by
CPI Group (UK) Ltd, Croydon CR0 4YY

Links to third party websites are provided by Oxford in good faith and
for information only. Oxford disclaims any responsibility for the materials
contained in any third party website referenced in this work.

# CONTENTS

| | |
|---|---|
| *Foreword by David Attenborough* | vii |
| *Introduction* | ix |
| *Acknowledgements* | xix |
| *The correspondents* | xxi |
| *List of letters* | xxvii |
| | |
| The Letters | I |
| England to Singapore and Malacca, 4 March–16 October 1854 (Letters 1–7) | 3 |
| Borneo, 17 October 1854–23 May 1856 (Letters 8–23) | 28 |
| Bali, Lombock, and Celebes, 23 May–16 December 1856 (Letters 24–31) | 84 |
| Aru and Amboyna, 17 December 1856–7 January 1858 (Letters 32–40) | 121 |
| Ternate and New Guinea, 8 January–8 October 1858 (Letters 41–49) | 151 |
| Batchian and Ternate, 9 October 1858–9 June 1859 (Letters 50–57) | 183 |
| Menado, Amboyna, and Ceram, 10 June 1859–16 June 1860 (Letters 58–67) | 207 |

CONTENTS

Waigiou, Ternate, and Timor, 17 June 1860–5 July 1861          227
    (Letters 68–75)
Java, Sumatra, and back home, 6 July 1861–31 March 1862          254
    (Letters 76–88)

*Appendix 1: Letter details*          286
*Appendix 2: Wallace's itinerary in the Malay Archipelago*          295
*Abbreviations*          303
*Illustration credits*          304
*Bibliography*          306
*Index*          309

# FOREWORD BY
# DAVID ATTENBOROUGH

These letters were written by one of the greatest of nineteenth-century naturalists, at one of the most crucial periods in the history of biology. That fact alone makes them fascinating reading. But they become the more gripping when one remembers that their author, Alfred Russel Wallace, was travelling in what was then one of the wildest and least explored parts of the world, the islands of Indonesia. Halfway through his journeying, as he lay shaking in the grip of a malarial fever, a theory came into his mind that would solve the mystery that was obsessing many zoologists of the time—how new species came into existence. After his fever subsided, he wrote an essay outlining his explanation and posted it to one of his regular scientific correspondents in Britain, Charles Darwin. Its arrival, for Darwin, was a profound shock for he himself had thought of exactly the same theory some 20 years earlier, but instead of publishing it he had spent his time since then accumulating evidence in support of it.

The problem of who should be regarded as the originator of the theory was famously solved by Darwin's scientific friends, who arranged for Wallace's essay and two extracts from Darwin's earlier manuscripts to be read out at the same meeting of the Linnean Society in London. In spite of this scrupulously even-handed arrangement, Darwin was generally perceived in London's scientific circles as the prime author of the theory and less than two years later that precedence was established by the publication of his magisterial book *The Origin of Species*.

Since then, some have claimed that Wallace was badly done by. A few have even suggested that Darwin falsified the date on which he received Wallace's essay in order to give him time to take some of Wallace's ideas and incorporate them into his own presentation. The editors of this book have been able to find evidence about postal dates that shows this accusation to be false. And Wallace himself had no doubt that Darwin was entitled to the greater credit because of all the work Darwin had done substantiating the theory. Nowhere in these letters does he show the slightest hint of resentment, whether he is writing directly to Darwin himself, or indeed to other more intimate naturalist friends to whom he might have felt free to express it. Indeed, the generosity of spirit that is evident on both sides is truly heartwarming.

The consequences of this fascinating episode echo throughout these letters, but there is much else in them to enjoy and relish. Writing to his close friend Henry Bates, with whom he had travelled in South America on an earlier expedition, Wallace reveals the competitive element that affects so many collectors by comparing the number of different species of beetles he has accumulated with the number found by Bates on the Amazon. Elsewhere—particularly when writing to his mother and sister—he gives more general accounts of his activities. Sometimes he recounts episodes in such detail that you can sense he has in his mind a bigger audience than his immediate correspondent. But anyone who has travelled in the remoter parts of Indonesia, particularly some 50 years ago or more, will need no reminding of what he seldom mentions—the arduous conditions in which he was working. Apart from a rare parcel of food from Europe, most of which had decayed into inedibility by the time it reached him, he lived entirely on local food. For much of the time he had no European companion. He had virtually no medicines except quinine with which he managed to subdue his malarial attacks. And he relied on small, flimsy local craft to take him on his frequent journeys from island to island. So these letters contain at one and the same time, the raw material, not only of the most influential of all biological theories, but of a thrilling story of exploration and take the reader into the mind of one of the most adventurous, observant, and honourable scientists of his time.

David Attenborough

# INTRODUCTION

Alfred Russel Wallace's voyage through the Malay Archipelago between 1854 and 1862 remains one of the most remembered and consequential of the Victorian era. Wallace was the first naturalist to collect on many of the islands he visited. He discovered thousands of new species and discovered that the fauna of the entire region was separated by a sharp division into Asian and Australian zones, which is still known as the Wallace Line. And in one of the most famous episodes in the history of science, Wallace formulated a theory of evolution by natural selection similar to that of Charles Darwin. Wallace later described his voyage as quite simply 'the central and controlling incident of my life'.[1]

This volume provides, for the first time, Wallace's surviving correspondence from that classic voyage. Letters were the prime means of communication allowing naturalists in the field to share their results with the scientific community and at the same time providing the field naturalist with the latest scientific news and publications. The mail service in the middle of the nineteenth century was incredibly efficient and relatively fast. The Peninsular & Oriental Company (P&O) operated a bimonthly mail steamer service between Southampton and Singapore. Letters and magazines from London could arrive in Singapore in just over six weeks. Of course, when Wallace moved to more remote islands in the Archipelago his access to mail became more intermittent, but in most places letters from home could reach him once every month or two.

---

[1] ML1: 335–6.

We have included here all of the letters to and from Wallace during his voyage in the Malay Archipelago that are known to survive. Many others, however, are lost, probably forever. Hopefully others may yet be found. Wallace seems to have written regularly to his mother and sister, perhaps six times a year at least. He probably received an even greater number in return. Instead of hundreds of family letters, we have just 22, and none written by his mother survive. In one of his letters to his friend George Silk (30 November 1858), Wallace wrote that he was sending more than a dozen letters in one month, but in this instance only three are known.

As so often with famous Victorians, many of Wallace's letters were published in his autobiography (1905) and in a collection after his death (1916). Scholars have relied on these for a century now. We have found that the older published versions had many omissions and sometimes alterations. All of the transcriptions in this collection have been made anew, directly from the original letters. For the first time, therefore, readers can access the complete texts of Wallace's first-hand accounts of his thoughts and adventures as originally written.

Wallace's classic two-volume account of his voyage, *The Malay Archipelago* (1869), is still in print. (See van Wyhe ed. 2015). His correspondence provides a fascinating alternative account of this famous voyage by Wallace himself—but one that is new to modern readers and contains many gems never seen before. The letters include many vivid descriptions of the beauty of the islands he explored, an appreciation of the rich biodiversity of the region, and provide fascinating ethnographic details of the many peoples he observed. Because many of his letters were intended for publication or to be read widely by family and friends, they are often written in an entertaining narrative. There are accounts of adventure—from man-eating tigers to shipwrecks and encounters with head-hunters—as well as Wallace's self-effacing manner with his mis-adventures humourously recounted.

The 1850s and 1860s were perhaps the height of the European obsession with collecting natural history specimens. Whereas collectors in previous eras favoured the rare, unusual, and particularly brilliant or spectacular types, in the mid-nineteenth century the focus shifted to compiling comprehensive collections of all the species of a certain group of animals or plants. Access to the more exotic and remote regions of the world was becoming easier with the advent of steamships and railways. In order to settle many of the important scientific questions of the day it was important to compare species from all parts of the world. Collectors in Britain were willing to pay decent prices, even for series of small insects. Some specimens were extremely valuable. Wallace chose to make his living by satisfying this lively demand for preserved creatures from the far corners of the globe. But he also hoped to publish his own master catalogue of the beetles of the Archipelago by retaining for himself a private collection to be fully studied after his return. He was a very successful collector, amassing (with the help of a large number of assistants) a total collection of 125,660 specimens among the groups that interested him most: insects, birds, and mammals.

Wallace travelled widely through the Malay Archipelago, first in Singapore, Malaysia, and Borneo, then on to the myriad islands of Indonesia and East Timor, and finally to Java and Sumatra. In *The Malay Archipelago* Wallace adopted a 'geographical, zoological, and ethnological arrangement' rather than a chronological one. He knew that most of the islands he visited would have been unknown to his readers, and his route was far from straightforward. This plan may have contributed to the lack of importance that exact dates seem to have had for him. We have found that very many of the dates Wallace provided were incorrect. We have reconstructed his entire itinerary using contemporary records and newspapers and particularly his unpublished notebooks, which often contain dates that Wallace himself overlooked when writing his book years later (see Appendix 2).

In February 1858 Wallace composed his historic essay (the 'Ternate' essay) on the origin of species and later sent it with a letter to Charles Darwin. It is one of the most famous and discussed episodes in the history of science. Some of the letters published for the first time in this volume throw new light on Wallace's famous voyage and on some of the questions that have been long debated in the scholarly and popular literature on Darwin and Wallace. For example, the letters reveal Wallace's previously unknown reluctance to publish openly on evolution. This suggests an answer to an old question: why did Wallace send his Ternate essay to Darwin rather than directly to a scientific journal for publication? There is also a fragment of a previously unrecognized Darwin letter copied into one of Wallace's notebooks, which adds details of Darwin's unpublished 'big book' on evolution that exist nowhere else. Other revelations offered by this volume include answers to two other old uncertainties. Who wrote first, Darwin or Wallace? And also of interest, which one first broached the topic of evolution in their correspondence? Contrary to what has long been believed, the answer is Darwin in both cases.

## BACKGROUND TO THE VOYAGE

Alfred Russel Wallace was born on 8 January 1823 near the market town of Usk in Monmouthshire (now Gwent) on the Welsh borders. He was the fifth child. Preceding him were two brothers, William and John, and two sisters, Elizabeth and Frances (known as Fanny). His parents, Thomas Vere Wallace (1771–1843) and Mary Anne (née Greenell, 1788–1868), were respectable gentlefolk who had recently fallen on financial hard times. Hence, the family could perhaps best be described as lower middle class. In 1828 the Wallace family moved to Hertford, north of London, where Wallace attended the gentlemen's Grammar School from 1829–1837. The school offered a classical education, not unlike Darwin's at Shrewsbury School, including Latin and English grammar, classical geography, writing, and mathematics. It was Wallace's only formal education. Although this is often described by modern writers as an unusual hardship, Wallace was in fact well-educated by the standards of the time. In Hertford, Wallace befriended a neighbour's boy named George Silk. The two remained friends for life and several letters from Wallace to Silk from the Malay Archipelago survive and are included here.

Wallace lived for a short time with his brother John in London in 1837. There, Wallace attended the working-class mechanics' institutes with their blend of self-help, religious scepticism, phrenology, and perhaps the socialism of Welsh reformer Robert Owen. Wallace became a religious freethinker and devoted to doctrines of a world controlled by natural laws rather than divine interventions. He then joined his eldest brother William as an apprentice land surveyor in the English midlands. The blend of outdoor work, surveying instruments, and mathematics pleased Wallace. During the following years of work in the countryside, he gradually took up an interest in natural history, first with botany as a hobby from 1841.

His father died in 1843 and was buried in the family vault in Hertford. At the same time demand for surveyors slackened. In early 1844 Wallace became a teacher at the Collegiate School in Leicester. There he met a new friend, the budding entomologist Henry Walter Bates. Bates introduced Wallace to beetle collecting. He was astonished to learn how numerous the species were, just around Leicester. In 1845 Wallace read an anonymous new book called *Vestiges of the natural history of creation* (1844).[2] The book convinced him of a type of progressive evolution—though it was very different from the theory that he and Darwin would later announce to the world in 1858. Indeed, Wallace seems to have largely adopted the theory of *Vestiges* until developing his own innovations between 1855 and 1858.

*Vestiges* argued that nature was controlled by a system of natural laws acting together to bring about progress. It was widely accepted that organisms produce offspring like themselves. Occasionally the action of some progressive natural laws would produce offspring of an even higher type than the parents. This ratchet effect led life to progress ever up the scale of being. Hence it appeared to make sense of the newly appreciated fossil record. By this time it was already widely accepted that the history of life had been generally progressive, since invertebrates appeared first, followed by fish, then reptiles, then birds, and finally mammals. Wallace was deeply impressed with this non-random order.

---

[2] Secord, J. 2000. *Victorian sensation: the extraordinary publication, reception, and secret authorship of* Vestiges of the natural history of creation. Chicago: University of Chicago Press.

In 1848 Wallace and Bates set out to seek their fortunes by turning their natural history hobby into a profession. They would travel to Brazil and collect natural history specimens along the Amazon and Rio Negro for sale on the lucrative collecting market in London. They engaged Samuel Stevens to act as their agent. Stevens had showrooms just across from the British Museum—one of his main customers.

In 1849 Wallace's sister Frances married Thomas Sims.[3] Sims was born in Swansea and by 1848 was one of the first commercial photographers in London using the daguerreotype process. Wallace had purchased some of the daguerreotype equipment during a brief trip to Paris in 1847. Wallace seems to have taken something of an authoritative attitude to his brother-in-law and many of the letters from the Malay Archipelago are addressed to him and often include paternal advice or discuss questions of commercial photography and optics.

After four years of collecting Wallace returned to Britain in 1852. Bates continued collecting on the Amazon. Tragically, however, the dilapidated ship on which Wallace had booked a passage caught fire and sank in the mid-Atlantic. Although rescued by a passing ship, Wallace lost most of his collection and notes. Fortunately, Stevens had insured Wallace's collection for £200. Nevertheless, most of Wallace's financial and scientific investment was lost.

Wallace set about giving talks on his findings at scientific meetings and publishing as much as he could from his few surviving records and from memory. He presented the Royal Geographical Society with a large map of the Rio Negro based on his own survey. All this networking succeeded in gaining Wallace enough clout to approach the eminent geologist Roderick Impey Murchison, until recently president of the Royal Geographical Society, for financial assistance for another collecting expedition.

An extraordinary lady traveller named Ida Laura Pfeiffer had recently visited the Malay Archipelago and her collecting was apparently known to Samuel

---

[3]   More on Thomas Sims and other correspondents of Wallace is contained in the section on Correspondents.

Stevens. The unusually high prices her specimens secured may have helped Wallace decide where to travel next. Singapore would be his destination. Eventually, financial assistance was secured from the government to buy Wallace a gentleman's first-class ticket on the P&O steamer route 'overland' via Egypt to Singapore. Wallace also took with him a young London boy, Charles Allen, as servant and assistant. Their combined first- and second-class tickets were £197—equal to the entire insured amount of Wallace's collection from Brazil. And so, on Saturday 4 March 1854 Wallace and Allen began their journey from Southampton on the paddle steamer *Euxine*. The first letter in this volume was written on this steamer.

## A NOTE ON THE POSTAL SYSTEM

Some considerable attention has been paid to the postal system for Wallace's correspondence, especially by Wallace enthusiasts John Langdon Brooks and Roy Davies.[4] However, these authors made many errors in their efforts to support accusations against Darwin of lying about the receipt of Wallace's letter and Ternate essay in 1858. Therefore, the literature on Wallace contains some incorrect details about the postal system that we hope to correct. For example, Brooks claimed that Wallace's Ternate essay could have reached Darwin (if mailed on 9 March) as early as 14–20 May 1858. But the earliest a letter posted on that date on Ternate via Southampton could reach London was 3 June. The route of the Dutch mails steamers through the Archipelago provided in Brooks' map was incorrect. Davies, knowing that Wallace's Ternate essay was sent in response to an earlier letter from Darwin that arrived on Ternate on 9 March 1858, imagined that Wallace 'must have opened it and read it at the quayside because of the reference to Lyell'.[5] Yet the Ternate steamer anchored in the bay

[4]  Brooks, J.L. 1984. *Just before the origin: Alfred Russel Wallace's theory of evolution.* New York and Guildford: Columbia University Press. Davies, R. 2008. *The Darwin conspiracy: origins of a scientific crime.* London: Golden Square books.
[5]  Davies, 2008, p. 142.

not at a quay, the mail was delivered to the post office in a sealed bag and not distributed to the public at the quayside, and the ship's crew were forbidden by contract to accept mail bound for Europe.

Wallace's correspondence to and from Britain travelled along two routes. The first was the P&O steamer line on which he travelled to Singapore. The P&O had a government contract to carry mail to India and Hong Kong.[6] European mail was received at Singapore twice a month. Mail bound for Europe was taken on a P&O steamer from Hong Kong. The steamer stopped at Penang and then Galle in Ceylon. From here the steamer would continue to Bombay. Mail for Europe was transferred to another ship that steamed from Galle to Aden and then Suez. Mail was transferred across the desert on camels until June 1854 when a rail line was opened between Cairo and Alexandria. The Suez Canal did not open until 1869. At Alexandria, the mail was transferred to another P&O steamer that stopped at the port of Valetta in Malta and then Gibraltar before arriving at Southampton. The mail reached London by rail at least by the following day. From there, transfer to other British cities could occur the same day by rail or perhaps the following day in the case of outlying towns such as Darwin's Downe in Kent.

There were two options using the P&O system. Letters marked 'via Southampton' travelled the route described above. A slightly faster and more expensive route was indicated as 'via Marseille'. This mail could be sent four days after the departure of the Southampton steamer and travel across the English channel by steamer, through France by rail and catch the Southampton steamer at Malta. The remainder of the route was the same. The transit from London to Singapore was therefore about 45 days via Southampton and 41 via Marseille. In 1855 a one-ounce letter from Singapore to London cost 1 rupee or $0.38. The post office in Singapore processed an impressive 400,000 letters each year.

The other way in which Wallace sent letters was with bulk items shipped to Britain via the Cape of Good Hope on sailing ships. This is how Wallace sent

---

[6]  Howarth, D. 1994, *The story of P&O: the Peninsular and Oriental Steam Navigation Company*. London: Weidenfeld and Nicolson.

almost all of his collections home to Stevens. This longer route took about 4–5 months from Singapore.

In the Dutch East Indies similar arrangements were available, combining mail and passenger services. A monthly steamer connected Singapore and Batavia, linking up with the P&O steamer bound for Europe. The Dutch paddle steamer, the *Koningin der Nederlanden*, was operated by the Nederlandsch-Indische Stoomboot Maatschappij (Dutch East-Indies Steamboat Company). In 1854, the Dutch government entered into a contract with the company owned by the former naval officer W.F.K. Cores de Vries to operate an additional service from Batavia to Singapore.[7] This increased the frequency of mail between Singapore and Batavia to bimonthly.

The mail service to eastern Java and the eastern islands of the Dutch East Indies was contracted to the Cores de Vries Company. There were four smaller screw steamers on this route. They performed a monthly circuit from Surabaya to Macassar, Coupang and Delli on Timor, Banda, Amboyna, Ternate, Menado, and again to Macassar. On return to Surabaya it connected with services to Batavia. Hence, at Wallace's most distant post office, Ternate, a letter could travel to or from Europe in about 77 days.

## EDITORIAL POLICY

We have sought to reproduce the original text of the letters faithfully following the original spelling, abbreviations, and punctuation. In most instances Wallace's handwriting is very clear and legible, but the legibility of his correspondents varies—with Sir James Brooke being the most difficult to decipher. Of course, every nuance of a manuscript document cannot be reproduced in print. For some scholars, therefore, recourse to the original may prove necessary. We have provided a free-flowing text, but one that preserves those elements normally

---

[7]   Anon. 1858. *Verslag van het beheer en den staat der Nederlandsche bezittingen.* Utrecht: Kemink, pp. 38-42. Campo, J.N.F.M. à. 2002. *Engines of empire.* Hilversum: Verloren, p. 41.

needed for scholarly citations. All the writing on the letters is transcribed, including address lines, dates, as well as postmarks.

We have adopted the following conventions in our transcriptions.

1. Text underlined in the original has been rendered as *italics*.
2. Text double underlined is printed as such.
3. ~~Deleted text~~ shows words deleted by the writer.
4. *[Some text]* is a conjectural reading of a hard-to-read word or passage.
5. Interlined text has been integrated in the main text, either where the writer indicated or where we have judged most appropriate.
6. Paragraph separations have been added where the writer left a larger space, long dash, or changed topics in a long block of text.
7. Annotations by recipients are recorded separately in a footnote.
8. Editorial annotations are given between [square brackets], and include English names for groups of animals identified by a scientific name, and persons indicated by initials or abbreviations.

This volume includes all surviving letters written by Wallace or received by him during his voyage in the Malay Archipelago, from March 1854 to March 1862. Besides letters, Wallace also sent home boxes with specimens and other items, and what today we would call journal articles meant to be published in one of the literary or scientific magazines. We have not included his manuscripts that were later published, even though these were sent as enclosures or in some instances have a locality and date included in them that resembles a letter. All papers authored by Wallace during his travels are reproduced in facsimile and full text forms on *Wallace Online* edited by John van Wyhe.

The letters are presented in chronological order. The full details of their current depositories and previous publication (often partially) are given in Appendix 2. Besides Wallace himself, this volume includes letters written to or by 15 persons. Short biographies of each are grouped together in one section, below on pp. xxi–xxvi.

# ACKNOWLEDGEMENTS

It is a pleasure to acknowledge the assistance of institutions and individuals for our research on this book. This work was conducted in conjunction with the development of the *Darwin Online* and *Wallace Online* projects at the Department of Biological Sciences at the National University of Singapore. An anonymous donor generously funded these projects during 2009–2013. We are especially grateful to David Attenborough for graciously contributing his foreword to this volume.

Judith Magee, George Beccaloni, and Caroline Catchpole of the Natural History Museum (NHM) in London were generous in providing information, assistance, and images from items in the museum collections. They conduct the Wallace Correspondence Project (WCP), which has become an unprecedented resource for future research on Wallace as well as the people and themes with which Wallace interacted during his long life. In turn we provided early transcripts of the letters in this volume to the WCP project to use as they saw fit. Since writing this volume they have launched *Wallace Letters Online* which includes facsimiles of the letters in this volume and much more.

At Oxford University Press, our Commissioning Editor, Latha Menon, was particularly helpful, encouraging and crucial in judiciously negotiating the unusually complex copyright and permissions pathways to make this book possible. We are thankful for the expertise shown by her team, particularly our Assistant Editor, Emma Marchant. We acknowledge the permission of the Trustees of the Alfred Russel Wallace Literary Estate to reproduce the texts of previously unpublished letters.

We are also grateful to Charles Allen's descendant Patricia Giudice, Gerrell M. Drawhorn, Anna Mayer, Christine Garwood, Adam Perkins, Tim Barnard, Jan Van Der Putten, Jonathan Hodge, Charles H. Smith, Andrew Wee, and Brenda Yeoh. At the Department of Biological Sciences we thank Paul Matsudaira, Lisa Lau Li-Cheng, Joan Choo Beng Goon, Yee Ngoh Chan, Nursyidah Binti Mansor, and Soong Beng Ching. At the Lee Kong Chian Natural History Museum thanks are due to Peter Ng and Leo Tan. Lucille Yap of the Philatelic Museum was helpful with several queries. We also thank the staff at the National Museum of Singapore and the National Archives of Singapore. At the Library of the Linnean Society of London Lynda Brooks, Gina Douglass, Leonie Berwick, and Andrea Deneau kindly gave us crucial assistance with the Wallace manuscripts in their collection. We acknowledge permission of the Syndics of Cambridge University Library to reproduce texts of documents in their collection.

Between 2003 and 2005 John van Wyhe was a research fellow on an earlier Wallace correspondence project at the Open University where he gained invaluable experience with Wallace's life and correspondence. That project, directed by James Moore, was largely modelled on the Darwin Correspondence Project at the University of Cambridge, which continues to serve as one of the authoritative models for scholarly correspondence projects. The experience of working on the Open University project was expanded into an unpublished book edited by van Wyhe of the correspondence of Johan Gaspar Spurzheim and later into the editorial work of *Darwin Online*. John van Wyhe also wishes to thank the Master, Fellows, and students of Tembusu College. Since completing this volume, the website *Wallace Letters Online* was launched which includes facsimiles of the letters in this volume and much more. We acknowledge the permission of the Trustees of the Alfred Russel Wallace Estate to reproduce the texts of previously unpublished letters.

# THE CORRESPONDENTS

**FREDERICK BATES** (1829–1903) was the younger brother of Henry Walter Bates, and as an amateur entomologist was particularly interested in *Heteromera*, a group of beetles. His general collection of this group, 22,930 specimens, was bought by Frederick Ducane Godman (1834–1919) and later transferred to the Natural History Museum, London. Frederick Bates founded the Leicester Brewing & Malting Co. (Upper Charnwood Street, Leicester), which operated the Eagle Brewery.

**HENRY WALTER BATES** (1825–1892) was the son of hosiery manufacturers in Leicester. He became a prominent entomologist. He went with Wallace to collect natural history specimens on the Amazon in 1848. While Wallace returned home in 1852, Bates stayed until 1859, having discovered thousands of new species of insects. From 1864 he worked as assistant secretary of the Royal Geographical Society in London, always maintaining his interest in the study of beetles.[8]

**SIR JAMES BROOKE** (1803–1868) was an English adventurer who became the hereditary Rajah of the district of Sarawak on Borneo who assisted and befriended Wallace.[9] He and Wallace saw each other again after Wallace's return to Britain.

---

[8] Crawforth, A. 2009. *The butterfly hunter: the life of Henry Walter Bates*, University of Buckingham Press.
[9] Runciman, S. 2010. *The white Rajah: a history of Sarawak from 1841 to 1946*. Cambridge University Press.

**CHARLES ROBERT DARWIN** (1809–1882) was a country gentleman and naturalist educated at Edinburgh and Christ's College, Cambridge. He joined the voyage of the *Beagle* commanded by Captain Robert FitzRoy (1805–1865) from 1831 to 1836 as naturalist. Darwin worked on his theory of evolution off and on for 20 years before publishing his famous book *On the Origin of Species* in November 1859.[10]

**JOHN GOULD** (1804–1881) trained as a gardener and taxidermist. He was curator of the new museum of the Zoological Society of London from 1827. He published several series of large, well-illustrated books on types of birds. After a visit to Australia in 1838–1840 he wrote books on the birds and mammals found there. He was one of the leading ornithological experts of his day. He also identified Darwin's bird specimens from the *Beagle* voyage.

**DANIEL HANBURY** (1825–1875) was a pharmacologist and botanist based in London. He investigated the medicinal properties and applications of plants.

**FRANÇOIS LOUIS NOMPAR DE CAUMONT LAPORTE, COMTE DE CASTELNAU** (1810–1880) studied natural history in Paris, travelled to the Americas, and from 1848 was in the French consular service. He was consul to Siam (Thailand) from 1858 to 1862. He collected and studied insects, fishes, and other animals.

**MARK MOSS** left Britain in 1838 to settle in Singapore, where he married a local girl called Maria. He appeared in the insolvents court in 1849 and was discharged because he had just served five-and-a-half years in prison.[11]

---

[10]   Wyhe, J. van. 2008. *Darwin.* London: Andre Deutsch. Wyhe, J. van. 2009. *Darwin in Cambridge.* Cambridge: Christ's College. Wyhe, J. van. 2009. *Charles Darwin's shorter publications 1829–1883.* Cambridge University Press. Chancellor, G. and Wyhe, J. van. 2009. *Darwin's notebooks from the voyage of the Beagle.* Cambridge University Press. All published work by Darwin and almost all of his manuscripts are available on the Internet: John van Wyhe, ed. 2002. *The complete work of Charles Darwin online*, <http://darwin-online.org.uk>.

[11]   Buckley, C.B. 1902. *An anecdotal history of old times in Singapore 1819–1867.* Singapore: Fraser & Neave, p. 503.

Together with William Waterworth, another commission agent, he sold two birds of paradise to Wallace in 1862.

**JAMES MOTLEY** (1822–1859) was a mining engineer and naturalist who went to Borneo to work in the coal mines which were then being established in Labuan. He left in 1853 and soon obtained a similar job in the Julia Hermina coal mine near Banjermassin, on the East coast of Borneo. He was much interested in all branches of natural history, and with L.L. Dillwyn published *The natural history of Labuan and the adjacent coasts of Borneo* in 1855.[12] He was murdered in the Banjermassin war in 1859 together with his wife and children, and all his possessions were looted or burnt.

**HENRY NORTON SHAW** (d. 1868) was a diplomat who served as secretary of the Royal Geographical Society, 1849–1863, and in 1866 became consul at St Croix in the West Indies.

**FRANCIS POLKINGHORNE PASCOE** (1813–1893) was a navy surgeon who devoted much of his time to entomology and came to specialize in beetles. He was president of the Royal Entomological Society between 1864 and 1865. His large collection of insects was incorporated in the Natural History Museum, London.

**PHILIP LUTLEY SCLATER** (1829–1913) was secretary of the Zoological Society in London from 1859. He was a keen ornithologist, with great interest in speciation and zoogeography. He reported on all additions to the Zoological Gardens of the Society in Regent's Park and was present at most of their meetings.[13]

---

[12] Denney, H. 1859. Biographical notice, with extracts from the correspondence, of the late Mr. Motley, who was massacred at Kalangan, May 1st, 1859. *Annals and Magazine of Natural History* (s3) 4: 313–317. Walker, A.R. 2005. James Motley (1822–1859): the life story of a collector and naturalist. *Minerva (Swansea History Society)* 13: 20–37.

[13] Brown Goode, G. 1896. *The published writings of Philip Lutley Sclater, 1844–1896.* Washington: Government Printing Office. Elliot, D.G. 1914. In memoriam: Philip Lutley Sclater. *Auk* 31 (1): 1–12.

In 1858 he proposed dividing the world into six zoological regions, which are still in use today. The famous 'Wallace Line' is a demarcation between two of these regions.

**GEORGE CHARLES SILK** (b. 1822) became one of Wallace's closest friends when they were living in Hertford. Silk was private secretary and reader to Archdeacon John Sinclair (1797–1875) until 1875. In 1901 he was listed in the UK census as a 'retired secretary' aged 78, living in Bloomsbury, London.

**FRANCES SIMS** (1812–1893) was Wallace's only surviving sister during his travels. She was called Fanny by her family. She was born on 24 June 1812 and christened on 17 July in St Marylebone Parish Church, London. On 15 February 1849, in Cadoxton-juxta-Neath, Wales, she married Thomas Sims, the photographer. Wallace frequently wrote to her about domestic news during his travels. Thomas and Fanny lived at 44 Upper Albany Street, near Regent's Park in London in 1852 (ML1: 263, 313), moving to Conduit Street in 1855. Fanny died on 14 September 1893, aged 81 years.

**THOMAS SIMS** (1826–1910) was the husband of Wallace's sister Fanny. Sims was an innovative early photographer initially using the daguerreotype process. Born in Swansea, he ran the Royal Park Photographic Studio in Upper Albany Street (February 1853 to July 1855) and in Conduit Street, off Regent Street (23 July 1855 to 1859). He then went to Tunbridge Wells and continued his photography business until the end of his life. Sims died at Tunbridge Wells on 14 November 1910.[14] Many of his papers are preserved in the Tunbridge Wells Museum.

**SAMUEL STEVENS** (1817–1899) was an entomologist. He became a member of the Entomological Society of London in 1837 and served as treasurer and member of its council. He worked in his father's J.C. Stevens Auction Rooms

---

[14] Ashton, E.R. 1930. Memoirs of a photographic pioneer [Thomas Sims]. *British Journal of Photography* (June): 353–5. Arnold, H.J.P. 1977. *William Henry Fox Talbot: pioneer of photography and science*. London: Hutchinson. Osman, C. 1991. Thomas Sims. *Photohistorian* 94: 82–3.

on King Street, Covent Garden, London with his brother John Grace Stevens from 1840. He opened his own Natural History Agency at 24 Bloomsbury Street, Bedford Square, London (across from the British Museum) in 1848. He acted as agent for Wallace's interests during his travels in Brazil and Southeast Asia.[15]

**JOHN WALLACE** (1818–1895) and **MARY WALLACE** (1834–1913). Alfred's brother John emigrated to California in 1849 at the time of the gold rush. In 1851 he joined the Tuolumne (or Columbia) Water Co. He married Mary Elizabeth Webster on 18 January 1855 and they had six children: John Herbert (2 August 1856–1914), William George (25 December 1857–1922), Mary Frances (18 October 1861–1934), Alfred Alexander (22 April 1864–1941), Percy Russell (1866–1917), and Arthur Henry (1868–1947). Alfred met John again in 1887 after 40 years during Wallace's lecture tour in America. John and his wife Mary settled in Stockton, California.[16]

**MARY ANN WALLACE** (1788–1868) was Wallace's mother. She was baptized on 18 June 1788 in St Andrew's Church, Hertford, the first daughter of John Greenell (1745–1824) and his first wife Martha, née Hudson. On 30 April 1807 Mary Ann married Thomas Vere Wallace (1771–1843), and they had nine children, of which Alfred Russel was the eighth. When Wallace left for the Malay Archipelago Mary Ann was living in 44 Upper Albany Street, London, and subsequently moved in with her daughter Frances (Fanny) Sims at 5 Westbourne Grove Terrace. She died on 15 November 1868 in Ifield, West Sussex.

---

[15] Baker, D.B. 2001. Alfred Russel Wallace's record of his consignments to Samuel Stevens, 1854–1861. *Zoologische Mededelingen* 75 (16): 251–341. Allingham, E.G. 1924. *A romance of the rostrum being the business life of Henry Stevens and the history of thirty-eight King Street, together with some account of famous sales held there during the last hundred years.* London: H.F. & G. Witherby. Carrington, J.T. 1899. Samuel Stevens. *Hardwicke's Science Gossip* (n.s.) 6: 161–2.

[16] Manna, S. 2006. Charles Darwin, Alfred Russel Wallace, evolution and the man after whom the Calaveras County town of Wallace is named. *Las Calaveras* July 2006. Eastman, B. 1969–1970. John Wallace and the Tuolumne Water Company. *Chispa* 9: 297–300, 308–311.

She probably wrote almost monthly to Wallace during his travels, but none of her letters to Wallace survive.

**GEORGE ROBERT WATERHOUSE** (1810–1888) was a naturalist and curator of the museum of the Zoological Society of London. He had a keen interest in entomology and with Frederick William Hope founded the Entomological Society of London in 1833, and was its president between 1849 and 1850.

# LIST OF LETTERS

## ENGLAND TO SINGAPORE AND MALACCA

1. Wallace to George Silk, 19 and 26 March 1854
2. Wallace to Mary Ann Wallace, 30 April 1854
3. Wallace to [Edward Newman], 9 May 1854
4. Wallace to Mary Ann Wallace, 28 May 1854
5. Wallace to Mary Ann Wallace, [24] July 1854
6. Wallace to Mary Ann Wallace, 30 September 1854
7. Wallace to George Silk, 15 October [1854]

## BORNEO

8. Wallace to Henry Norton Shaw (RGS), 1 November 1854
9. Wallace to Samuel Stevens, [c. December 1854]
10. Wallace to Samuel Stevens, 8 April 1855
11. Wallace to John Wallace, 20 April 1855
12. Wallace to George Robert Waterhouse, 8 May 1855
13. Wallace to Frances Sims, 25 June 1855
14. Wallace to Frances Sims, 28 September 1855 and 17 October 1855
15. Wallace to Mary Ann Wallace, 25 December 1855
16. Charles Darwin to Wallace, [December 1855]

17. Wallace to Frances Sims, 20 February [1856]

18. Wallace to Thomas Sims, [c. 20 February 1856]

19. Wallace to Samuel Stevens, 10 March 1856

20. Wallace to Frances Sims, 21 April 1856

21. Wallace to [Frances and Thomas Sims], [21 April 1856]

22. Wallace to Henry Walter Bates, 30 April and 10 May 1856

23. Wallace to Samuel Stevens, 12 May 1856

# BALI, LOMBOCK, AND CELEBES

24. James Brooke to Wallace, 4 July 1856

25. Wallace to Samuel Stevens, 21 August 1856

26. Wallace to Samuel Stevens, 27 September 1856

27. James Brooke to Wallace, 5 November 1856

28. Henry Walter Bates to Wallace, 19 and 23 November 1856

29. Wallace to Samuel Stevens, 1 December 1856

30. Wallace to John and Mary Wallace, 6 December 1856

31. Wallace to Frances Sims, 10 December 1856

# ARU AND AMBOYNA

32. Wallace to Samuel Stevens, 10 March and 15 May 1857

33. Charles Darwin to Wallace, 1 May 1857

34. Wallace to Henry Norton Shaw, August 1857

35. Wallace to Charles Darwin, [27 September 1857]

36. James Brooke to Wallace, 31 October 1857

37. Wallace to Samuel Stevens, 20 December 1857

38. Charles Darwin to Wallace, 22 December 1857

39. Wallace to Philip Lutley Sclater, [January 1858]

40. Wallace to Henry Walter Bates, 4 and 25 January 1858


## TERNATE AND NEW GUINEA

41. Wallace to Frederick Bates, 2 March 1858

42. James Motley to Wallace, 22 May 1858

43. Charles Darwin to Wallace, [13 July 1858]

44. Wallace to Samuel Stevens, 2 September 1858

45. Wallace to Henry Norton Shaw, September 1858

46. Wallace to Frances and Thomas Sims, 6 September 1858

47. Wallace to [Samuel Stevens], 5 October 1858

48. Wallace to Mary Ann Wallace, 6 October 1858

49. Wallace to Joseph Dalton Hooker, 6 October 1858


## BATCHIAN AND TERNATE

50. Wallace to Samuel Stevens, 29 October 1858

51. Wallace to Francis Polkingthorne Pascoe, 28 November 1858

52. Wallace to George Silk, 30 November 1858

53. Charles Darwin to Wallace, 25 January 1859

54. Wallace to Samuel Stevens, 28 January 1859

55. Charles Darwin to Wallace, 6 April 1859

56. Wallace to Thomas Sims, 25 April 1859

57. Daniel Hanbury to Wallace, 6 May 1859

# MENADO, AMBOYNA, AND CERAM

58. Wallace to Francis Polkingthorne Pascoe, 20 July 1859

59. Charles Darwin to Wallace, 9 August 1859

60. Wallace to John Gould, 30 September 1859

61. Wallace to Philip Lutley Sclater, 22 October 1859

62. Charles Darwin to Wallace, 13 November 1859

63. Wallace to Henry Walter Bates, 25 November 1859

64. Wallace to Samuel Stevens, 26 November 1859 and 31 December 1859

65. Wallace to Samuel Stevens, 14 February 1860

66. Daniel Hanbury to Wallace, 17 February 1860

67. Charles Darwin to Wallace, 18 May 1860

# WAIGIOU, TERNATE, AND TIMOR

68. Wallace to George Silk, 1 September 1860 and 2 January 1861

69. Wallace to Charles Darwin, [December 1860]

70. Wallace to Samuel Stevens, 7 December [1860]

71. Wallace to Philip Lutley Sclater, 10 December 1860

72. Wallace to Francis Polkingthorne Pascoe, 20 December 1860

73. Wallace to Henry Walter Bates, 24 December 1860

74. Wallace to Samuel Stevens, 6 February 1861

75. Wallace to Thomas Sims, 15 March 1861

# JAVA, SUMATRA, AND BACK HOME

76. Wallace to Mary Ann Wallace, 20 July 1861

77. François de Caumont LaPorte to Wallace, 26 August 1861

78. Wallace to Philip Lutley Sclater, [11–15] and 20 September 1861

79. Wallace to Frances Sims, 10 October 1861

80. Wallace to Charles Darwin, 30 November 1861

81. Wallace to Henry Walter Bates, 10 December 1861

82. Wallace to George Silk, 22 December 1861 and 20 January 1862

83. Mark Moss to Wallace, 3 February 1862

84. Mark Moss to Wallace, 6 February 1862

85. Wallace to Philip Lutley Sclater, 7 February 1862

86. Wallace to Samuel Stevens 16 February 1862

87. Wallace to Philip Lutley Sclater, 18 March 1862

88. Wallace to Philip Lutley Sclater, 31 March 1862

# THE LETTERS

# THE LETTERS

## ENGLAND TO SINGAPORE AND MALACCA
### 4 March–16 October 1854

Wallace left England on 4 March 1854 on the P&O paddle steamer *Euxine*. He apparently wrote only one letter during the voyage, to his boyhood friend George Charles Silk. The letter was posted mid-voyage at Aden. Wallace's letters home were then shared with his mother, sister, and brother-in-law, and some were apparently copied and sent to his brother John in California. In this first letter, Wallace writes about life aboard a passenger steamer and his impressions of chaotic but exciting Egypt. In passing he mentions his assistant, Charles Martin Allen, who was just 14 years old.[17] Wallace would teach Allen how to shoot and skin birds and to catch and preserve insects. We know from his letters that Wallace was quite dissatisfied with Allen, whom he found untidy and slow to improve. No letters from Allen are known to survive but he presumably found working for Wallace rather difficult. After about two years, Allen found different employment in Sarawak, but was again engaged by Wallace as a freelance collector from January 1860 to February 1862 (see Appendix 2).

---

[17] Charles Martin Allen (1839–1892) came out to the Malay Archipelago in the company of Wallace in 1854. Allen worked for Wallace (1854–1856) and later in the Moluccas and New Guinea (1860–1862). Afterwards he lived in Singapore and had a variety of jobs, finally becoming the manager of the Perseverance Estate. See Rookmaaker, K. and van Wyhe, J. 2012.

Fig. 1. The paddle steamer *Euxine*

After Aden, Wallace and Allen sailed on a second steamer, the *Bengal*, to Ceylon. The *Bengal* was bound for India so Wallace and Allen changed to the P&O line bound for Hong Kong via Singapore. They arrived in Singapore aboard the P&O paddle steamer *Pottinger* on 18 April 1854, although for some reason Wallace always gave the date of arrival as the 20th.[18] After staying a week in a hotel, they moved eight miles inland to stay with the Rev. Anatole Mauduit, a French Catholic missionary in the Bukit Timah district. The low hills surrounding the mission, such as Bukit Timah and Bukit Batok, were still clothed in virgin forest, often cut through with roads and footpaths that provided Wallace's first hunting ground for natural history specimens in Asia.

Their next collecting location was also British: the town and territory of Malacca on the Malay Peninsula to the north. They sailed on 13 July 1854, arriving in Malacca two days later. Wallace probably had a letter of introduction because we know he spent two days with another French missionary in the town. They soon moved further inland to Gading, 13 miles from the town, where again Wallace wrote to his mother. Soon after, they made an excursion to the top of nearby Mt Ophir, where they were expecting to see elephants, rhinoceroses, and maybe a tiger, but in fact encountered none.

---

[18] *Straits Times*, 19 April 1854, p. 9.

4

Returning to Singapore at the end of September, Wallace and Allen apparently stayed again at Bukit Timah. Wallace called on the Rajah of Sarawak, Sir James Brooke, who was in Singapore to attend a commission to investigate allegations that he had employed excessive force in suppressing pirates. Sir James offered to assist Wallace on a visit to his territory of Sarawak, Borneo. This decided Wallace's next destination, and interestingly it was also in effect another British territory. After two further letters home, Wallace and Allen set sail for Sarawak.

# Singapore

## 1. To George Silk, 19 and 26 March 1854

### Mediterranean

*March 19, 1854*

All right so far.[19] We have had beautiful weather. I was sick only one day and qualmish two or three more. This ship is crowded, we have four berths in our cabin, 3 occupied & there is not room for two to dress in it at once. I am in the lower tier, they are luckily well ventilated and so are pretty comfortable. We sit down 60 to dinner, everything is generally cold & the only way to get a good dinner is to seize on the nearest dish to you and stick to it till exhausted nature is replenished. Our hours are Breakfast 9 Lunch 12, Soda water Ale & bread & cheese wine & spirits ad lib. Dinner at 4 claret port & sherry with champagne twice a week. Tea at 7 (Band plays) 8 to 9. Lights extinguished at ½ past 10.

---

[19]   The top of the letter is annotated: 'Copy of Alfreds Letter Euxine Steamer to G. Silk'.

Fig. 2. George Silk

Our company consists of a few officers and about 20 cadets for India, 3 or 4 Scotch clerks for Calcutta, same number of business men for Australia. A Chinese interpreter[20] and two or three others for China, Frenchmen. A Portuguese officer for Goa with whom I converse, 3 Spaniards going to the ~~Phillip~~ Philippines (very grave) a gentleman and two Ladies (Dutch) going to Batavia and some officers & miscellaneous for Alexandria & others we have left at Malta. There are two or three chess players & I ~~some~~ have so much improved that I think I should have some chance with you. At Gibraltar we lay 24 hours in Quarantine. A boat came alongside with a health officer in the stern, selling fruit, cigars &c, all handed up with tongs and payment received in a basin of water, all for fear of the cholera.

We reached Malta at night & left at 10 the next morning. I went on shore from 6 to 9, saw the streets and the market, heard maltese spoken, admired the beggar boys & girls and walked through the cathedral of St. John gorgeous with gold and marble and the tombs of the Knights of Malta. We have had lovely weather but the boat is slow going generally only 9 Knots, so we are two days behind time, but we hope to make it up on the other side of the Isthmus in the "Bengal" which is a fast boat and much more commodious having 2200 tons & the cabins much longer. Tomorrow morning we are to reach Alexandria & then for Cairo and the desert with a glimpse of the Pyramids I hope. A parson came

[20]    ML1: 332 has: 'A Chinese Government interpreter'.

on board at Malta going to Jerusalem, Mr. Hayward. My namesake also came on board there, he goes to Bombay, where he has been before. He is a neat figure, sharp face and very respectable, not at all like me![21]

I have found no acquaintance on board who exactly suits me, so shall have less to regret at parting with them. One of my cabin mates is going to Australia, reads "How to make money", seems to be always thinking of it and is very dull and unsociable, the other is a young cadet, very aristocratic great in Dressing case & Jewellery.[22] Takes an hour to dress & reads the Hindostany Grammar. The frenchman, the Portuguese & the Scotchman I find the most amusing; there is a little fat navy Lieut. who is very amusing & tries practical jokes and has set up a "Monté" Table.[23] If you have no more than this it will go from Alexandria. If not I will continue it. *So Adieu.*

## Steamer "Bengal" Red Sea
### *March 26th*

I now go on with my account and hope to give *you some idea of the state of Egypt &c &c*. Of all the eventful days of my life my first in Alexandria was the most striking. I imagine my feelings when coming out of the Hotel (whither I had been convey'd in an omnibus) for the purpose of taking a quiet stroll through the City, I found myself in the midst of a vast crowd of donkey's & their drivers all thoroughly determined to appropriate my person to their own use and interest, without in the least consulting my inclinations. In vain with rapid strides and waving arms I endeavored to clear a way and move forward, arms and legs

---

[21]  This could be the person listed as Mr. L.A. Wallace in *Allen's Indian Mail*, 20 March 1854, p. 147.

[22]  Possibly Freedley, Edwin T. 1853. *How to make money: a practical treatise on money; with inquiry into the chances of success and causes of failure.* London.

[23]  Monté, also known as three-card monte, is a card game in which a victim is supposed to select the money card after the set has been shuffled on a table or box.

were seized upon, and even the Christian coat-tails were not sacred from the profane Mahometans.

One would hold together two donkeys by their tails while I was struggling between them, & another, forcing together their heads, would thus hope to compel me to mount upon one or both of them; and ~~another forcing~~ one fellow more impudent than the rest I laid flat upon the ground, and sending the donkey staggering after him, I escaped a moment midst hideous yells and most unearthly cries. I now beckoned to a fellow more sensible-looking than the rest, and told him that I wished to walk and would take him for a guide, & hoped now to be at rest; but vain thought! I was in the hands of the Philistines, and getting us up against a wall, they formed an impenetrable phalanx of men and Brutes thoroughly determined that I should only get away from the spot on the legs of a donkey.

Bethinking myself now that donkey-riding was a national institution, and seeing a fat Yankee (very like my Paris friend)²⁴ mounted, being like myself hopeless of any other means of escape, I seized upon a bridle in hopes that I should then be left in peace. But this was the signal for a more furious onset, for seeing that I would at length ride, each one was determined that he alone should profit by the transaction, and a dozen animals were forced suddenly upon me and a dozen hands tried to lift me upon their respective beasts. But now my patience was exhausted, so keeping firm hold of the bridle I had first taken with one hand, I hit right & left with the other, and calling upon my guide to do the same, we succeeded in clearing a little space around us. Now then behold your friend mounted upon a jackass in the streets of Alexandria, a boy behind holding by his tail and whipping him up. Charles (who had been lost sight of in the crowd) upon another, and my guide upon a third, and off we go among a crowd of Jews and Greeks, Turks and Arabs, and veiled women and yelling donkey-boys to see the City.

---

²⁴ This may be a reference to the 'stout, good-humoured American, a New-England manufacturer, going to Paris on business for the first time' who accompanied Wallace and Silk on their travels from London to Dover and on to Paris in 1853, when they were on their way to Switzerland (ML1: 325).

We saw the Bazars & the slave market, where I was again nearly pulled to pieces for "backsheesh" (money), the Mosques with their elegant minarets, and then the Pasha's new Palace, the interior of which is most gorgeous. We have [seen] lots of Turkish soldiers walking in comfortable irregularity; and, after feeling ourselves to be dreadful Guys for two hours, returned to the hotel whence we were to start for the canal boats. You may think this account is exaggerated, but it is not; the pertinacity, vigour and screams of the Alexandrian donkey-drivers no description can do justice to.—on our way we passed Pompey's Pillar & then ~~crowded~~ a day rowed in a small boat on a canal, the next day on the Nile, mud villages, Palm trees, camels & irrigating wheels turned by [blank] buffaloes form the staple of the landscape with a perfectly flat country often beautifully green with crops of corn & lentils, numerous boats with immense triangular sails.—

Here the Pyramids came in sight, looking very large, then a handsome castellated bridge for the Alexandria & Cairo railway—and then Cairo, Grand Cairo! the city of romance which we reached just before sunset. We took a guide & walked in the city, very picturesque, and dirty, got to a quiet English hotel where a mussulman waiter rejoicing in the name of "*Ali Baba*" gave me some splendid Tea, brown bread & fresh butter. One or two French and English travellers were the only company & I could hardly realize my situation. I longed for you to enjoy it with me. Thackeray's *first day in the East* is admirable, read it again, it represents my sentiments exactly.[25] In the morning at 7 we started for Suez in small two wheeled omnibuses 6 in each 4 horses change every 5 miles, a meal every 3 hours at very comfortable stations, the desert is undulating, covered with a coarse volcanic gravel. The road is excellent, hundreds of camels skeletons lay all along, vultures & a few sand grouse were seen, also some small sand larks &c. We saw the mirage frequently. Near the middle station the Pasha has a hunting lodge like a palace.

---

[25]  [Thackeray, W.M.] 1846. *Notes of a journey from Cornhill to Grand Cairo*. London: Chapman.

The Indian & Australian mail about 600 boxes & all the parcels & passengers luggage came over on camels which we passed on the way—a few odoriferous little plants grew here & there in the hollows, I made a small collection in my pocket books and got a few landshells. We enjoyed the ride exceedingly and reached Suez at midnight.—Suez is a miserable little town & the Bazar is extraordinarily small, dark & dirty, no water or any green thing exists within ten miles. In the afternoon we were taken on board the ship, a splendid vessel where the cabins are large & comfortable and everything very superior to the Euxine.[26] It is a perfect calm & at length hot & pleasant.

I cannot write on board ship or I could have filled sheets with an account of "*From Alexandria to Suez*" I was much pleased. Take this letter to Nº 44 Albany Street[27] as I cannot now write home.—

Yours sincerely

Alfred R. Wallace.

G.C. Silk, Esq.

## 2. To Mary Ann Wallace, 30 April 1854

**Singapore**

*April 30th. 1854*

My dear Mother[28]

We arrived here safe on the 20th of this month, having had very fine weather all the voyage.[29] On shore I was obliged to go to a Hotel which was very

---

[26] This 'splendid vessel' was the much larger and more luxurious *Bengal*. Built in Glasgow only two years before at a cost of £68,300, she could carry 135 first-class passengers and had a crew of 115.

[27] The London address of Wallace's mother, where she lived together with her daughter Frances and Thomas Sims.

[28] This letter is endorsed '2nd letter' in the top left corner of the first sheet.

[29] The date is incorrect. Wallace arrived on the P&O steamer *Pottinger* on 18 April 1854, cf. *Straits Times*, 19 April 1854, p. 9.

expensive,[30] so I tried to get out into the country as soon as I could, which however I did not manage in less than a week, when I at last got permission to stay with a French Roman Catholic Missionary who lives about 8 miles out of the Town & close to the Jungle.[31] The greater part of the Inhabitants of Singapore are Chineese, many of whom are very rich, & all the villages about are almost entirely of Chinese, who cultivate pepper & Gambir.[32] Some of the English merchants here have splendid country houses. I dined with one to whom I brought an introduction. His house was most elegant & full of magnificent Chineese & Japan furniture.[33]

We are now at the Mission of *Bukit Tima*.[34] The missionary speaks English, Malay & Chineese, as well as French, and is a very pleasant man. He has built a pretty church here & has about three hundred Chineese converts. Having only been here four days, I cannot tell much about my collections yet. Insects however are plentiful. Some sorts more so than at Para, others less. I have not yet decided what I shall do but I think to stay here a month & then go to Malacca till my boxes arrive by the 'Eliza Thornton'.[35]

The mail from England will be in tomorrow or next day when I shall expect some letters & papers. I wrote last to George Silk from Aden & told him to

[30]  Wallace and Allen probably stayed at Singapore's London Hotel, renamed in 1865 as Hotel de l'Europe.
[31]  The missionary, never named by Wallace, was Rev. Anatole Mauduit (1817–1858). The Church of St Joseph remains, although the building Wallace saw was replaced in the 1960s with the present one.
[32]  Gambier was used for dying.
[33]  The name of this English merchant is unknown. It was likely either George Garden Nicol (1810–1897) or John Jarvie, the Singapore partners of Hamilton, Gray & Co., whose names appear at the top of Wallace's address list in *Notebook 4*.
[34]  Bukit Timah is Singapore's highest hill. It's name means 'tin hill' in Malay.
[35]  The barque *Eliza Thornton* left London on 15 March and arrived in Singapore on 12 July 1854, cf. *Straits Times*, 18 July 1854, p. 8.

show you the letter. Letters addressed to me, care of Mess^rs. Hamilton Gray will reach me, till further orders.[36]

Tell Mr. [Samuel] Stevens I will write to him next mail & tell him a little about the Insects &c. I have not seen any tigers yet & do not expect to, for there are not many in this neighbourhood & there has not been a man killed at the place for two years.

Charles gets on pretty well. He is quite well in health & catches a few insects; but he is very untidy, which you may imagine by his clothes being all torn to pieces by the time we arrived here. He will no doubt improve & will soon be useful. Singapore is a very curious & interesting place. The Chineese do all the work, they are a most industrious people, & the place could hardly exist without them. The harbour is full of Chinese junks & small native vessels. There are now also 3 English men of war as some Russian vessels were expected here.

Malay is the universal language, in which all business is carried on. It is easy, & I am beginning to pick up a little, but when we go to Malacca shall learn it most, as there they speak nothing else. I am very unfortunate with my watch. I dropped it on board & broke the balance spring, & have now sent it home to Mr. Matthews[37] to repair, as I cannot trust anyone here to do it. The bill will be sent to Mr Stevens who please tell to pay it. There was also a book I left unpaid for 10^s/6^d I think at Williams & Norgate.[38] Tell Mr. S. to pay this also. Every

---

[36] Wallace's agent in Singapore was Hamilton, Gray & Co., established there in 1832 as a branch of the Scottish firm of Buchanan, Hamilton & Co. The *Singapore almanack and directory* for 1854 (p. 44) listed the partners as Walter Buchanan and William Hamilton (England), George Garden Nicol and John Jarvie (Singapore). The office was located on 'Left Bank, River Side'. Assistants and clerks were Edward Lose, William Kraal, R.S.S. Padday, F.H. Marcus, and Boon Wat.

[37] Wallace stayed with William Matthews of Leighton Buzzard for some months in 1838 to learn about the trade of watch- and clock-making. Matthews moved to London (10 Bunhill Row, St Luke's) soon afterwards. See MLI: 135–9.

[38] Williams & Norgate were booksellers and publishers who specialized in foreign scientific literature, at Henrietta Street, Covent Garden, London.

Fig. 3. Wallace with his mother, and his sister Fanny standing

thing for living in the Town of Singapore is very dear except clothing &c. I have had a lot of jackets & trousers made by a Chineese very well at 2s. each for making.

Love to Fanny & Thomas.[39]

I remain your affectionate Son
Alfred R. Wallace.

P.S. The climate here is almost exactly that of *Para* & is very delightful. The forest too looks very much the same. Palms are very abundant. AW.

---

[39] Frances (Fanny) Wallace (1812–1893) and her husband Thomas Sims (1826–1910).

Address—"A.R. Wallace, Esq$^r$ to the care of Mess$^{rs}$ Hamilton Gray & C$^o$. Singapore (via Southampton)[40]

---

## 3. To [Edward Newman], 9 May 1854

### Singapore

*May 9, 1854*

As I have no doubt that my entomological friends will be glad to hear that I have arrived safe, and have commenced work, I will give you a short account of my progress up to this time.

I landed at Singapore on the 20th of April, after a 46 days' passage from England without any incident out of the common. For a week I was obliged to remain in the town at an hotel, not finding it easy to obtain any residence or lodging in the country. During this time I examined the suburbs, and soon came to the conclusion that it was impossible to do anything there in the way of insects, for the virgin forests have been entirely cleared away for four or five miles round (scarcely a tree being left), and plantations of nutmeg and Oreca palm have been formed.[41] These are intersected by straight and dusty roads; and waste places are covered with a vegetation of shrubby Melastonias, which do not seem attractive to insects. A few species of Terias, Cethosia, Danais and Euploea, with some obscure Satyridæ [brush-footed butterflies], are the only butterflies seen, while two or three lamellicorn beetles on the Acacia trees were the only Coleoptera [beetles] that I could meet with.

At length, however, I obtained permission to reside a few weeks at a Roman Catholic mission near the centre of the island, from which place, called 'Bukit Tima,' I now write. Here portions of the forest, which originally covered the

---

[40]   There were two mail options for sending letters to Singapore. The first 'via Southampton' followed the steamer route that Wallace undertook to Singapore. The other option 'via Marseille' was more expensive and allowed letters posted four days later to travel across France to meet up with the P&O steamer at Malta. The remainder of the route was the same.

[41]   Oreca, rather Areca palms, have nuts which are chewed with betel leaves for mild stimulating effects.

whole island, and which is rapidly disappearing, still exists, and it is in them that I find my only good hunting-grounds.

From the highest point in the island near here (only 500 feet) a good view is obtained of the plantations which are everywhere formed by the Chinese for the cultivation of pepper and gambic [gambier]; and it is apparent that but few years can elapse before the whole island will be denuded of its indigenous vegetation, when its climate will no doubt be materially altered (probably for the worse), and countless tribes of interesting insects become extinct.[42] I am therefore working hard at the insects alone for the present, and will give you some little notion of what I have done and may hope to do.

First, then, in Lepidoptera [moths and butterflies] I have been tolerably successful, having in about twelve days obtained 80 species of Diurnes [true butterflies]. If other localities prove equally rich I think the Eastern Archipelago may not fall much short of S. America. I have already about 30 species of Lycænidæ [gossamer-winged butterflies] and Ericinidæ [metalmark butterflies], some of which I have no doubt will prove new. Among the larger species the most remarkable is a magnificent Idæa, which is abundant in the forest, sailing or rather floating along, and having to my eye a far more striking and majestic appearance than even the Morphos of Brazil. It was a great treat to me to behold them for the first time, as well as many other of the Eastern forms to which I had pretty well familiarised my eye in collections at home. The Euplœas here quite take the place of the Heliconidæ of the Amazons, and exactly resemble them in their habits. I have taken the singular Danais Daos, *Doub.* [Doubleday], figured by Boisduval[43] as an Idæa, which it exactly resembles in its colour,

---

42   The administration believed that deforestation of the hilltops would adversely affect the climate of the island, as expressed by the solicitor and journalist James Richardson Logan, see Logan, J.R. 1848. The probable effects on the climate of Pinang of the continued destruction of its hill jungles. *Journal of the Indian Archipelago and Eastern Asia* 2: 534–6.

43   Jean Baptiste Alphonse Dechauffour de Boisduval (1799–1879), French lepidopterist and physician. As part of the scientific results of the voyage of the *Astrolabe*, he published *Faune entomologique de l'Océanie* in 2 volumes:

markings and flight; indeed there are small specimens of the Idæa from which it cannot be distinguished till captured, yet it is certainly a true Danais. The Leptocercus Curius is not uncommon here; it is a Papilio Protesilaus in miniature. Of true Papilios I have only four common species, and one of the group resembling Euplœa, which may prove new.

I must now turn to the Coleoptera. I am delighted with them; for though all small at present, they are exceedingly beautiful and interesting. I have 6 species of Cicindelas [tiger beetles], all small; 13 Carabidæ [ground beetles], mostly minute, but very beautiful; 10–12 Cleridæ [chequered beetles]; about 30 very small Curculionidæ [weevils]; and, *mirabile dictu!* 50 species of Longicornes [long-horned beetles], and it is only ten days since I took the first. Imagine my delight at taking 8 to 10 a day of this beautiful group, and almost all different species; but the worst of it is that I have got into a place where there are many woodmen and sawyers at work, and it is in the neighborhood of the fallen timber that I get most of them, on the wing. Almost all are small, few exceeding an inch and many not much more than a line. Under Boleti I have found some extraordinary Erotylidæ [fungus beetles]. The Elaters [click beetles] and Buprestidæ [wood-boring beetles] are all very small, as well as the Chrysomelidæ [leaf beetles] and other small groups. In all I have now 250 species, which will increase daily, but at a slower rate.

In the other orders there is nothing very remarkable. Hemiptera [bugs], as well as bees and wasps, are very scarce. Tenthredinidæ [sawflies] are rather abundant. Of dragon-flies I have many pretty species, and the Diptera [flies] are plentiful and very curious. I have taken a species (of the genus Diopsis I believe) with telescopic eyes, and some other singular forms. Ants are very abundant, also scorpions and centipedes, but these I do not seek after.

In the midst of this entomological banquet there is, however, one drawback—a sword suspended by a hair over the head of the unfortunate flycatcher: it is the possibility of being eaten up by a tiger! While watching with eager eyes some

*Lepidoptères* (1832) and *Coléoptères, Hémiptères, Orthoptères Névroptères, Hyménoptères et Diptères* (1835).

lovely insect, the thought will occasionally occur that a hungry tiger may be lurking in that dense jungle immediately behind intent upon catching you. Hundreds of Chinamen are annually devoured.[44] Pitfalls are made for the animals all over the country; and in one of them, within two miles of our house, a tiger was captured a short time before my arrival. Only last night a party of Chinamen, going home to their plantation, turned back afraid, having heard the roaring of a tiger in the path. These are unpleasant reminders of the proximity of a deadly foe; and though perhaps the absolute danger is little enough, as the tiger is a great coward and will not attack unless he can do it unawares, yet it is better to have the mind quite free from any such apprehensions. I shall therefore most probably leave here in a month or so for Borneo, before which, however, I hope to make such a collection as to give a tolerably correct idea of the Entomology of Singapore.

## 4. To Mary Ann Wallace, 28 May 1854

### Bukit Tima, Singapore

*May 28th. 1854*

My Dear Mother

I send you a few lines through G. S. [George Silk] as I thought you would like to hear from me. I am very comfortable here living with a Roman Catholic missionary, a frenchman who speaks Chineese, Malay & English. I & Charles go into the Jungle every day for insects, we have seen no tigers in fact they are getting scarcer every day as the jungle is more cleared away. I shall probably stay here another month & then go to Borneo. I send by this mail a small box of insects for Mr. Stevens. I think a very valuable one & I hope it will go safely.

---

44   Wallace was repeating a very popular legend about Singapore, that tigers killed about one person per day. Twenty-one deaths were reported in the newspapers in 1855. Chua, K. 2007. The tiger and the theodolite: George Coleman's dream of extinction. *Focas* 6: 124–67.

I expected a letter from you by the last mail, but received only two Athenaeums of March 18. & 25. Did not you send the 11th & April 2nd?——

I expected to receive some invoice of the goods I sent by the "Eliza Thornton". Did not you receive any? Tell Thomas I wish he would send me a copy of each of the pictures he took of me in the next letter. Address as before. I hope there is plenty of business. I do not think there is a Daguerrotypist in Singapore but it would not I think answer for long as living is very expensive. I saw a report of Mr. Wilsons paper at the Geolog. Soc. in the Athenaeum.[45] Is he gone yet? Send John's letters or a copy of them to me.[46]

The forest here is very similar to that of S. America. Palms are very numerous, but they are generally small & horridly spiny. There are none of the large & majestic species so abundant on the Amazon. I am so busy with insects now that I have no time for anything else. I send now about a thousand beetles to Mr. Stevens, & I have as many other insects still on hand which will form part of my next & principal consignment. Singapore is very rich in beetles & before I leave I think I shall have a most beautiful collection.

I will tell you how my day is now occupied. Get up at half past 5. Bath & coffee. Sit down to arrange & put away my insects of the day before, & set them safe out to dry. Charles mending nets, filling pincushions & getting ready for the day. Breakfast at 8. Out to the jungle at 9. We have to walk up a steep hill to get to it & always arrive dripping with perspiration. Then we wander about till two or three, generally returning with about 50 or 60 beetles, some very rare & beautiful.[47] Bathe, change clothes & sit down to kill & pin insects. Charles do [ditto] with flies, bugs & wasps, I do not trust him yet with beetles.

---

[45] Wallace presumably refers to Wilson, J.S. 1854. On the gold regions of California. *Quarterly Journal of the Geological Society* 10: 308–21. The paper was read to the Society on 8 March 1854.

[46] John Wallace (1818–1895), Alfred's brother, who lived in California.

[47] In ML1: 338 this sentence reads: 'Then we wander about the delightful shade along paths made by the Chinese wood-cutters till two or three in the afternoon, generally returning with about 50 or 60 beetles, some very rare & beautiful and perhaps a few butterflies'.

Dinner at 4. Then at work again till six. Coffee—Read—if very numerous work at insects till 8-9, then to bed—Adieu, with love to all.

Your affectionate son Alfred R. Wallace.

Mrs. Wallace
44 Upper Albany St.
fac^d by G. C. *Silk*, Esq.

# Malacca

## 5. To Mary Ann Wallace, [24] July 1854
### In the Jungle near Malacca
*July, 1854*

My dear Mother[48]

As this letter may be delayed getting to Singapore, I write at once having an opportunity of sending to Malacca tomorrow. We have been here a week living in a Chineese house or shed, which reminds me remarkably of my old Rio Negro habitations. I have now for the first time brought my "rede"[49] into use, and find it very comfortable.

We came from ~~Malacca~~ Singapore in a small schooner with about 50 chineese, hindoos & portuguese passengers, & were two days on the voyage, with nothing but rice & curry to eat, not having made any provision, it being our first

---

[48] The letter is annotated '5th letter'. Wallace only dated his letter as 'July, 1854' but added that he had been a week in the place where he wrote it. He reached the village of Gading outside Malacca on 18 July, and a week from then would be 24 July.

[49] Brazilian Portuguese for hammock.

experience of this country vessels. Malacca is an old Dutch city, but the Portuguese have left the strongest mark of their possessions in the common language of the place being still theirs. I have now two Portuguese servants, a cook & a hunter & find myself thus almost brought back again to Brazil by the similarity of the language, the people, & the jungle life. In Malacca we staid only two days, being anxious to get into the country as soon as possible. I staid with a Roman Catholic Missionary, there are several here, each devoted to a particular part of the population, Portuguese, Chineese & wild Malays of the Jungle.[50] The gentleman we were with is building a large church, of which he is architect himself, & superintends the laying of every brick & the cutting of every piece of timber. Money enough could not be raised here, so he took a voyage *round the World!* and in the United States, California, & India got subscriptions sufficient to complete it.

It is a curious & not very creditable thing that in the English Colonies of Singapore & Malacca there is not a single protestant missionary; while the conversion, education & physical & moral improvement of the inhabitants (non-European) is entirely left to these French Missionaries, who without the slightest assistance from our Government devote their lives to the christianising & civilising the varied populations which we rule over.

Here the birds are abundant & most beautiful, more so than on the Amazon, & I think I shall soon form a most beautiful collection. They are however almost all common, & so are of little value except that I hope they will be better specimens than usually come to England. My guns are both *very good*, but I find powder & shot in Singapore *cheaper* than in London, so I need not have troubled myself to take any. So far both I & Charles have enjoyed excellent health. He can now shoot pretty well & is so fond of it that I can hardly get him to do anything else. He will soon be very useful, if I can cure him of his incorrigible

---

[50]   Wallace stayed with Pierre Etienne Lazare Favre (1812–1887) who was assigned to the Chinese Catholics at Malacca. St Francis Xavier's Church was built between 1849 and 1856. Decroix, P. 2005. *History of the Church and churches in Malaysia and Singapore (1511–2000)*. Penang, pp. 225–6. Wallace recorded the barometer reading of 'Mr Favrex' at the front of *Notebook 1* on 16 July 1854.

carelessness. At present I cannot trust him to do the smallest thing without watching that he does it properly, so that I might generally as well do it myself.

The Chineese are most industrious. They clear & cultivate the ground with a neatness which I have never before seen in the tropics, & save every particle of manure both from men & animals to enrich the ground. The country about Malacca is much more beautiful than Singapore, it being an old settlement with abundance of old fruit & forest trees scattered about. Provisions are cheaper, but every thing else dearer than in Singapore. Monkeys of many sorts are abundant here, in fact all animal life seems more plentiful than in Brazil. Among fruits I miss most the oranges, which are scarce & not good; & there is no substitute for them.

I shall remain here probably 2 months & then return to Singapore to prepare for a voyage to Cambodia or somewhere else, so do not be alarmed if you do not hear from me regularly. Love to all at home from

Your ever affectionate Son

Alfred R. Wallace.

Mrs Wallace

# Singapore

## 6. To Mary Ann Wallace, 30 September 1854

Sept.$^r$ 30th. 1854

*Singapore*

My dear Mother

I last wrote to you from Malacca in July. I have now just returned to Singapore after two months hard work.

At Malacca I had a pretty strong touch of fever with the old Rio Negro symptoms, but the Government doctor made me take a great quantity of quinine every day for a week together & so killed it, & in less than a fortnight I was quite well & off to the Jungle again. I see now how to treat the fever, & shall commence at once when the symptoms again appear. I never took half enough quinine in America to cure me.

Malacca is a pretty place, & I worked very hard. Insects are not very abundant there, still by perseverance I got a good number & many rare ones. Of birds too I made a good collection. I went to the celebrated Mount Ophir and ascended to the top.[51] The walk was terrible 30 miles through jungle, a succession of mud holes. My *boots* did good service.

We lived there a week at the foot of the mountain in a little hut built by our men, & I got some fine new butterflies there and hundreds of other new and rare insects. We had only rice & a little fish & tea, but came home quite well. The height of the mountain is about 4,000 feet. Near the top are beautiful ferns & several kinds of fir trees of which I made a small collection. Elephants & Rhinoceroses as well as Tigers are abundant there, but we had our usual bad luck in not seeing any of them.

On returning to Malacca I found the accumulations of two or three posts, a dozen letters & fifty newspapers—my watch & pins &c from Mr Stevens. I had letters from Algernon & my Uncle.[52] The latter wants me to visit Adelaide, the former wants to visit me but is afraid he cannot manage it. Your letter contained much news. Mr [Thomas] Sims come to London is a miracle. You do not say whether any of you have been to the Crystal Palace yet. Even G.S. [George Silk] who admires nothing says it is indescribable.

---

[51] The most conspicuous landmark visible from Malacca is Mount Ophir (Gunung Ledang). There were several gold mines on the mountain.

[52] Algernon Wilson (1818–1884), Wallace's cousin, was Registrar of Probates and an enthusiastic amateur entomologist living in Adelaide, Australia (ML2). See: Wilson, J.G. 2005. *The insect man: Charles Algernon Wilson: Adelaide's first entomologist.* Henley Beach: S. Aust. Seaview Press. The person referred to as 'my Uncle' was Algernon's father and husband of Wallace's maternal aunt Martha, Thomas Wilson (1787–1863), a solicitor and author who emigrated to Australia in 1838 and settled in Adelaide.

I am glad to be safe in Singapore with my collections, as from here they can be insured. I have now a fortnight's work to arrange, examine, & pack them, & then in four months hence there will be some work for Mr. Stevens.

Sir James Brooke is here. I have called on him. He received me most cordially, & offered me every assistance at Sarawak. I shall go there next, as the Missionary does not go to Cambodia for some months. Besides, I shall have some pleasant society at Sarawak, & shall get on in Malay, which is very easy, but I have had no practice—though still I can ask for most common things.

My books & instruments arrived in beautiful condition. They looked as if they had been packed up but a day. Not so the unfortunate eatables. We were all very stupid to pack them up in a basket at all. Nothing but tin cases will preserve such things. The pudding & twelfth cake were masses of mould & insects, quite useless. The covers of the jars were all eaten through by ants and small insects. The currant jam was mostly spoilt, *sour*. The gooseberry remained very good. Anything of the sort put into tin cases & soldered up, which would not cost more than 6$^d$ & would no doubt arrive perfectly good.

I shall probably have a box sent in a few months then you can try the experiment. I am sorry you did not send my splayed shoes in Mr Stevens' parcel. The shirts do not send on any account, I have too many here at present.

The butterfly I sent in a letter I knew was not rare. I merely sent it as a specimen of a kind which there is nothing resembling in America.

If it were not for the expense, I would send Charles home. I think I could not have chanced upon a more untidy or careless boy. After 5 months I have still to tell him to put things away after he has been using them as the first week. He is very strong & able to do any thing, but can be trusted to do nothing out of my sight.

Please put a stamp on the enclosed letter & post it.

With Love to all I remain

Your affectionate Son

Alfred R. Wallace.

Mrs. *Wallace*

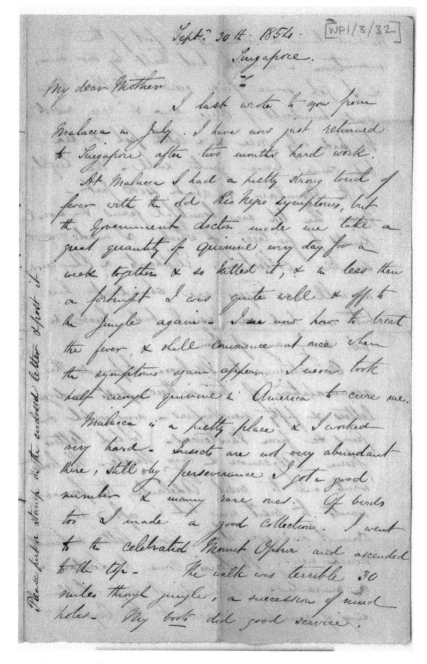

Fig. 4a. Wallace's letter to his mother, Mary Ann Wallace, 30 Sept 1854

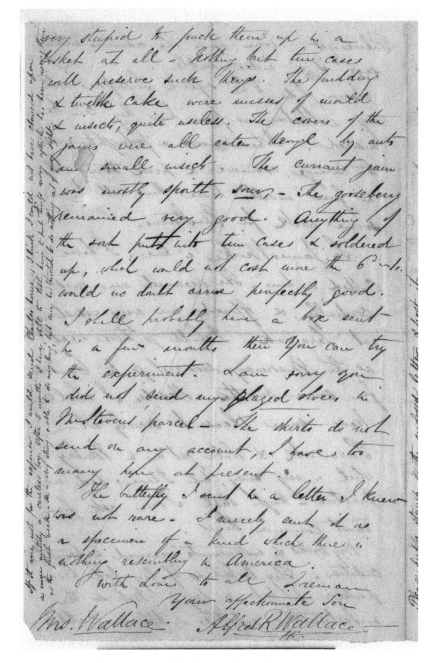

Fig. 4b. Second part of Wallace's letter to his mother, 30 Sept 1854

## 7. To George Silk, 15 October [1854]

**Singapore**
*Oct. 15<sup>th</sup>*

Dear G.

To morrow I sail for Sarawak. Sir J. Brooke has given me a letter to his nephew, Capt<sup>n</sup> Brooke[53] to make me *at home* till he arrives, which may be a month perhaps. I look forward with much interest to see what he has done & how he governs. I look forward to spending a very pleasant time at Sarawak. These four sheets are for the Lit. Gaz.—correct ad lib. I think you need not rewrite a copy.[54] The mail is due tomorrow but may not come in time for me, as I leave at 12.

I have had very hard work for 3 weeks here packing up, arranging & cataloguing all my collections; about 6000 specimens of Insects, Birds, ~~Animals~~ Quadrupeds & shells. If too long this letter may be cut in two, at the 2nd page of sheet 2, where I have used a cross would do for a division. I am too hard worked to write home now, so please call at *44* & tell them, that they are not to expect to hear from me for *three mails* now, as communications from Sarawak are not very regular.

Sir W. Hooker's[55] remarks are encouraging, but I ~~have~~ cannot afford to collect plants. I have to work for a living, & plants would not pay unless I collect nothing else, which I cannot do, being too much interested in Zoology. I should like a botanical companion like Mr. Spruce very much.[56] We are anxiously expecting accounts of the taking of Sebastopol.

---

[53] Captain John Brooke (1823–1868), nephew and heir-apparent of Sir James Brooke.

[54] Wallace, A.R. 1855. Extracts of a letter From Mr. Wallace. *Hooker's Journal of Botany* 7 (7): 200–9 [S19].

[55] William Jackson Hooker (1785–1865), botanist and the first director of the Royal Botanic Gardens at Kew in 1841.

[56] Richard Spruce (1817–1893), botanist. Wallace and Spruce became friends on the Amazon in 1849. After Spruce's death, Wallace edited his *Notes of a botanist on the Amazon and Andes*. London: Macmillan & Co. 2 vols, 1908.

I am much obliged to Latham[57] for quoting me & hope to see it soon. That ought to make my name a little known. I have not your talent at making acquaintances & find Singapore very dull. I have not found a single agreeable companion. I long for you to walk about with & observe the queer things in the streets of Singapore. The Chinamen & their ways are inexhaustibly amusing. My revolver is too heavy for daily use. I wish I had had a small one.

Yours sincerely

Alfred R. Wallace.

P.S. A Punch[58] now & then will be very acceptable. Mr Stevens will be sending a box soon.

Excuse the awful uncorrectedness of the scrawl.

G.C. *Silk*, Esq.

[Enclosure][59]

our men however declared one day they had seen a rhinoceros. We heard the fine Argus pheasants every evening, but they were so wild that it was impossible to get a sight of them. Our rice being finished, & our boxes crammed full of specimens, we returned, our men taking us by what they termed a better road, winding about through Malay villages, & making our second day's walk upwards of 30 miles. I only stayed at Ayer Panas, sufficient time to pack up all my collections, & then returned to Malacca on my way to Singapore. We were congratulated by all our friends on having lived a week at the foot of Mt Ophir without getting fever.—A.R.W.

---

[57] Wallace heard about a summary of his South American travels contained in Latham, Robert G. 1854. *The natural history department of the Crystal Palace described: Ethnology*. London: Crystal Palace Library, Bradbury & Evans, pp. 64–8.

[58] The satirical London magazine *Punch*, established in 1841.

[59] On the reverse of the letter to Silk there is a fragment of text that is identical to the end of a paper communicated to Hooker's *Journal of Botany*, dated 10 October 1854. This must be part of the enclosure mentioned in the letter to Silk.

# BORNEO
## 17 October 1854–23 May 1856

No contemporary accounts survive of Wallace's voyage to the territory of Sarawak on the great island of Borneo, 500 miles east of Singapore. Wallace and Allen probably set out on the brig *Weraff* on 17 October 1854 and arrived at Kuching, the capital of Sarawak, on 29 October.[60] Wallace would spend well over a year collecting in various locations in western Borneo, his longest continuous stay on any island. At the end of 1854 Wallace wrote a short essay about his first impressions of Sarawak, which was published anonymously in the *Literary Gazette* of 9 June 1855.[61] He had been staying 'some time at a cottage of Sir James Brooke's, about twenty miles inland, on the ridge of a mountain, at an elevation of about one thousand feet. The path up is peculiar, half is over broken rocks, the other half up ladders'. The bungalow on Serambu Hill was called Peninjau (lookout).[62] There were three similar routes into the mountains leading to three villages of the aboriginal Dyak tribes. Wallace had been to see their longhouses built on stilts, and recorded details of their dress and way of life. The local gold mines were worked by Chinese, who lived in neat villages. Wallace had hardly started to collect the local fauna when the wet season put a halt to full-day collecting.

---

[60] *Straits Times*, 17 October 1854, p. 8 and 24 October 1854, p. 8. Wallace mentioned his arrival in a letter to Norton Shaw of 1 November 1854, hence the date given in MA1: 54 (1 November 1854) must be erroneous. In fact this was the day Wallace started collecting.

[61] [Wallace, A.R.] 1855. Sarawak, Borneo, 1854. *Literary Gazette and Journal of the Belles Lettres, Science, and Art* No. 2003 (9 June): 366 [S18].

[62] Wallace used 'Peninjauh' to refer to the mountain rather than to the bungalow (MA1: 128, 131). Long-term resident Spenser St. John (St. John, S. 1863. Vol. 1, p. 166) stated that Peninjau meant 'look-out'.

As the rainy season continued in February 1855, Wallace had time to write his now famous Sarawak Law paper on the succession of species during the history of life on earth.[63] After surveying the geographical distribution of animals and plants, and acknowledging the geological age of the Earth, he formulates a bold conclusion that *'Every species has come into existence coincident both in space and time with a pre-existing closely allied species'*.[64] Although this is commonly interpreted as his first published declaration of his belief in evolution, he in fact nowhere suggests that new species actually descended from earlier ones. Instead, he states only that new species were somehow 'created' or 'formed' according to earlier models or antitypes and in the same region. Hence, there were two likely ways to read the paper at the time. First, as a way of showing how the paleontological and geographical evidence is consistent with an evolutionary theory or, secondly, according to a natural law new species were created in coincident times and places. In later life Wallace himself, like modern writers, wrote about his paper as if it were a discussion of evolution. While still in the Archipelago however, Wallace referred to this paper not as one on evolution, but as 'On Law of Succession of Species'.[65]

When the rains finally stopped, Wallace was ready to explore the riches of the Bornean fauna further afield. On 14 March, Wallace and Allen travelled north by ship to a tributary of the Sadong River called Si Munjon, where a new coal works was being opened under the direction of an English mining engineer named Robert Coulson. Wallace usually referred to the area as Borneo rather than Sarawak, as it was only recently ceded to Sir James Brooke by the Sultan of Brunei, and not part of the territory of Sarawak. While staying in a bamboo house at Si Munjon, Wallace made daily excursions into the forest to collect

---

[63] Wallace, A.R. 1855. On the law which has regulated the introduction of new species. *Annals and Magazine of Natural History* (ser. 2) 16 (93): 184–96 [S20].

[64] Italics in the original.

[65] *Notebook 4*, p. 122; Wallace to Darwin [27 September 1857] 'my views on the order of succession of species' and Wallace to H. Bates 4 January 1858 'my paper on the succession of species'. See Wyhe, J. van. 2013. *Dispelling the darkness*.

insects and birds. His daily routine is nicely described in his letter to Samuel Stevens of 8 April 1855. Wallace hoped to clear up conflicting observations in the scientific literature of the number of different species of orang-utan that lived in the area. In his letter to his sister Fanny of 25 June 1855 he amusingly refers to his little orphan baby, which will have startled her until she realized that he was talking about a small orang-utan. While housebound due to an infected ankle, Wallace continued to work on his evolutionary speculations in his notebooks. While at Si Munjon he noted in particular the arguments against evolution in his copy of Charles Lyell's *Principles of geology* (1835).

After some eight months at Si Munjon, Wallace returned overland along the Sadong River to Sarawak. At this time Charles Darwin was seeking information about domesticated breeds of animals from around the world for his species theory.[66] He sent a series of enquiries to about 30 correspondents, including Wallace, via Samuel Stevens.[67] As Darwin's original letter does not survive, most writers on Wallace have been unaware of its existence. However, a memorandum preserved in the Darwin Archive in Cambridge served as a model, or aide memoir, for Darwin's letters, and we have included it here as the closest approximation of what the first letter between the two men contained, approximately dated to December 1855.

But of even greater significance is a comparison of the three letters to other Westerners on Darwin's list that still survive.[68] These letters are all extremely similar in sequence and content so we can be reasonably confident that the letter to Wallace was comparable. The opening line probably explained how

---

[66]  Secord, J. 1981. Nature's fancy: Charles Darwin and the breeding of pigeons. *Isis* 72: 162–86.

[67]  Wallace recalled that the letter came through Stevens. Wallace, A.R. 1887. [Letter to Alfred Newton, dated 3 December 1887]. In: Darwin, F. 1892. *Charles Darwin: his life told in an autobiographical chapter.* London: John Murray, pp. 189–90.

[68]  Darwin's letters to Edgar L. Layard, 9 December 1855 (CCD5: 524); to George H.K. Thwaites, 10 December 1855 (CCD5: 526); and to Charles A. Murray, 24 December 1855 (CCD5: 530).

Darwin came by Wallace's contact details and then expressed Darwin's familiarity with Wallace's travels and publications. Next, Darwin mentioned his ongoing research on the origin of species. That Darwin declared this to Wallace in their very first communication has not, to our knowledge, been previously noticed. The wording in all three letters is very similar. For example, Darwin wrote to the British envoy to the court of Persia, C.A. Murray, that 'I have for many years been working on the perplexed subject of the origin of varieties & species, & for this purpose I am endeavouring to study the effects of domestication, & am collecting the skins of all the smaller domesticated birds & quadrupeds from all parts of the world'. As a result of this enquiry, Wallace sent a 'domestic duck' to Darwin in a shipment from Lombock in 1856 (see letter to Samuel Stevens, 21 August 1856). We can only imagine what Wallace must have said in any note or letter sent in reply—but given that Darwin had broached the subject of evolution, and Wallace's deep fascination with the subject, it seems plausible that Wallace may have mentioned the subject again.

At some time during his stay Wallace captured a tiger beetle 'singularly agreeing in colour with the white sand of Sarawak' (to Frederick Bates, 2 March 1858). Tiger beetles are fast running predatory beetles that often inhabit open ground. This was the first of a series of differently coloured tiger beetles that Wallace collected on different islands, closely matching the colour of their habitat.

According to the *Malay Archipelago*, Wallace left Sarawak on 25 January 1856. But shipping records and his correspondence show that he must have departed on the bark *Santubong* on 10 February and arrived in Singapore on 17 February.[69] Charles Allen remained in Sarawak, having decided to work at the local Anglican mission. Instead, Wallace brought his 'Malay boy' Ali with him, a native of Sarawak, then about 15 years old.[70] He served Wallace at first as a cook and

---

[69]   *Straits Times*, 19 February 1856, p. 8.
[70]   ML1: 382. Ali was born in Sarawak, and married in Ternate when staying there with Wallace. Fichman, M. 2004. *An elusive Victorian: the evolution of Alfred Russel Wallace*. Chicago: University of Chicago Press, pp. 33–4.

servant, and gradually became a collecting assistant and eventually Wallace's chief assistant. The two would remain almost always together until the end of the voyage in 1862. After returning from Borneo, Wallace remained in Singapore waiting for a ship that would take him to Macassar on Celebes until 23 May 1856. During this time Wallace wrote letters to his family, to Samuel Stevens, as well as his first letter from the Archipelago to his friend Henry Walter Bates, who was still collecting in the Amazon.

# Sarawak

## 8. To Henry Norton Shaw (RGS), 1 November 1854

**Sarawak**

*Nov*. 1st. 1854*

My dear Sir,[71]

Having now reached my destination, & being about to commence work, I take the opportunity of communicating the fact through you to the Geographical Society, & also to give a short account of the places I have already visited.

I spent the first two months after my arrival, in the island of Singapore and a small island between it & the main land [Pulau Ubin], and made an extensive collection of insects, a great proportion of which are new to science & of very great interest to the Entomologist. I then went to Malacca, where I also spent

---

[71] Wallace's letter of 1 November 1854 was read in the meeting of the Royal Geographical Society on 26 February 1855. In the presidential 'Address to the Royal Geographical Society of London; delivered at the Anniversary meeting on the 28th May, 1855' by the Earl of Ellesmere (*Journal of the Royal Geographical Society of London* 25), receipt of the letter is acknowledged (p. xcvi) followed by an abstract of the contents (p. cxv).

two months visiting several localities in the interior & making an excursion to Mount Ophir, which I ascended, & by means of careful observation with Adie's Sympiesometer ascertained to be 3920 feet above the sea.—It is an isolated mountain, & in fact there appears to be no connected chain in this part of the peninsula. The summit is almost pure quartz, below more or less granitic & at the base I found highly inclined stratified rocks of a crystalline sandstone. Between it & the sea coast the country is ~~very flat, but with~~ low, the base of the mountain not being more than 200 feet above the sea. The Physical Geography of the province of Malacca is interesting & I will make a few observations on it.

Large tracts of land adjoining the coast were perfectly flat, & scarcely elevated above the sea. They are swamps and are cultivated as paddy fields. Low undulating hills of the curious volcanic conglomerate called *laterite* rise out of these flats; several of these occur near the Town of Malacca & give the appearance from the sea of an elevated country. On passing them inland however, we saw that they are but islands rising out of a swamp. These occur at intervals for some miles inland, where ~~del~~ spurs or points of the more elevated country project into the level country, ~~from~~ leaving wide flat valleys between them. A few miles further & these valleys have all contracted to a few hundred yards wide & ~~wind~~ wind about between low undulating banks, appearing exactly like the beds of large rivers. These flat valleys penetrate quite up to the base of Mt. Ophir & the other mountains in the interior of the peninsula, where a tract of low flat country seems to connect them

Fig. 5. Henry Norton Shaw

with others, ~~flowing falling to~~ having an outlet on the E. side of the peninsula, there appearing to be *no dividing ridge between them.*

Charts of the Straits of Malacca show ~~there~~ it to be a continuation below the sea of the same kind of country, the water being very shallow for a great distance from the shore & the islands which rise out of it, corresponding to the hills which rise from the paddy fields of the peninsula. The whole country is a dense jungle,—there appear to be no tracts of naturally open ground.

At Malacca I collected insects extensively, also land shells, birds &c. & obtained an acquaintance with the inhabitants, the scenery, & the animals & vegetable productions of this ~~E.~~ portion of the East, which will be of great value in my exploration of the less known country I have now reached.

On returning to Singapore I found my books & instruments had arrived safe from England. Sir James Brooke was there, ~~attending the commission of inquiry now sitting~~ & as he most kindly offered me every assistance in exploring the territories under his rule, I determined to come here at once.

I must not forget to mention that I forwarded the letter from the Dutch Government to Batavia & have received a reply stating that I should meet with no obstructions in visiting any of their Eastern possessions.

I am much pleased with the appearance of the country (though I am only three days here) there being many more hills than I expected offering facilities for *mapping*, which I trust to make a good use of.

I remain,

Yours faithfully,
Alfred R Wallace

D<sup>r</sup>. Norton Shaw.[72]

---

[72] Annotation below the text of the letter: 'Dr Norton Shaw. Sec. Geog. Soc.—At the suggestion of the Royal Geographical Society Mr Wallace after his return from South America, was kindly provided by the Earl of Clarendon with a free passage to Singapore and with letters of recommendation from the Governments of Holland and to the Governors of Java and the Philippine Islands'.

Fig. 6. Samuel Stevens

## 9. To Samuel Stevens, [c. December 1854]

Mr. Stevens read an extract from a letter[73] received from Mr. Wallace, in Borneo, in which he stated that he had taken about 700 Micro-Lepidoptera, among which are some extraordinary developments of palpi, &c. He finds these small moths come in abundance to a lamp, on dark, wet nights, and in the wet season he is sure he could get thousands of them.

# Si Munjon, Borneo

## 10. To Samuel Stevens, 8 April 1855

### Si Munjon Coal Works, Borneo

*8th April, 1855*

You will see by the heading of this letter that I have changed my locality. I am now up the river Sadong, about twenty miles N.E. of Sarawak. A small coal-field has

---

[73] The original of this letter read in the meeting of the Entomological Society of London on 2 April 1855 has not been found. As it was written in Borneo and reference is made to the wet season, it was probably written at the end of 1854.

been discovered here, and is now being worked. At present the jungle is being cleared, and a road made to carry the coals to the river side, and it is on account of the scarcity of roads in this country that I thought it advisable to come here. Another reason was, that this is the district of the 'Mias' or Orang-utan, the natural history of which I am very anxious to investigate, so as to determine definitely whether or no three species exist here, and also to learn something of their habits in a state of Nature. An English mining engineer has the direction of the works here, and has about a hundred Chinese labourers engaged.[74] I am residing with him, at the foot of the hill in which the coal is found. The country all round us is dead level and a perfect swamp, the soil being a vegetable mud, quite soft, and two or three feet deep, or perhaps much more. In such a jungle it is impossible to walk; a temporary path has, however, been made from the river (about a mile and a half) by laying down trunks of trees longitudinally.

Along this path is very good collecting-ground, but many fine insects are daily lost, and butterflies can hardly be captured at all, from the impossibility of stepping out of the path, and the necessity of caution in one's movements to preserve balance and prevent slipping, not at all compatible with the capture of active tropical insects. The small clearing in which our houses are situated also furnishes me with many good insects among the trunks and stumps, and other decaying timber. Half a mile further on in the jungle, on the hill side, is another clearing, where coal levels are to be opened; and, lastly, the jungle is being cut down to form a road or railway, and which, as it progresses, I expect will offer me very fine collecting-grounds.

Having now been here nearly a month I can offer some opinion on its entomological capabilities. Imagine my delight in again meeting with many of my Singapore friends,—beautiful longicorns of the genera Astathes, Glenea and Clytus, the elegant Anthribidæ [fungus weevils], the pretty little Pericallus and Colliuris, and many other interesting insects. But my pleasure was increased as

---

74 The English mining engineer was Robert Coulson, who had worked for the Eastern Archipelago Company in Labuan in 1851 and later settled in Singapore.

I daily got numbers of species, and many genera which I had not met with before. Longicorns I think are more abundant than at Singapore, and more than half of them are new to me. The species, too, run a size larger. Some of the scarcest there are here the most abundant, while many of the commonest of that locality seem to be altogether absent from this. Curculionidæ [weevils] are about equal in number, and there is a fair proportion of novelty. Two or three species closely allied to the Mecocerus Gazella are abundant, and many curious Brenthidæ [straight-snouted weevils].

I am paying much attention to the most minute species, and can safely promise abundance of novelty for Mr. [George] Waterhouse [at the British Museum]. Carabidæ [ground beetles] are hardly so abundant as at Singapore, but I have some beautiful new Therates, Catascopus and Colliuris, and the curious Thyreoptera also occurs sparingly under Boleti. The Cleridæ [checkered beetles] seem very similar to those of Singapore, but scarcely so numerous. The Buprestidæ [wood-boring beetles] I am happy to say are very fine; not that the species run very large, but they are tolerably abundant. One of the most beautiful I make out to be Belionota sumptuosa, about an inch long, and of the richest golden copper-colour; it flies with the greatest rapidity, making a loud bee-like hum, and settles on timber only in the hottest sunshine. There are also many smaller species of a dark green, variously marked with lighter green or golden spots (Chrysobothris sp.?). Also several slower flying kinds, which when fresh are clothed with a yellow powder, like the Euchroma gigantea? of Brazil, which, however, seldom reaches England in that condition, as it is difficult to capture and kill the insect without injuring its delicate covering. I have also some very curious minute species, making altogether thirty-six species of this interesting family collected in Borneo. I also pay much attention to the Elateridæ [click beetles], and have many pretty things, especially among the velvety species, with a swollen thorax. Of Heteromera, Erotylidæ, Chrysomelidæ and Trimera, I have hosts of curious things, which are daily increasing in number.

The only family in which there is an absolute deficiency, is that of the beautiful Cetoniadæ [rose chafers]. I have only at present one or two Tænioderas, a fine green and black Coryphocera, and the handsome Macronota Diardi, which is, I believe, very rare: I scarcely dare hope to increase my collection of this family to any great extent, as they evidently are only abundant in mountainous and rather open shrub-producing districts, while they are scarcely at all represented in the dense and gloomy jungles which are the favourite haunts of all those insects which at any period of their existence feed on fresh or decaying timber, or on the boleti which grow upon it. Among my latest captures are, my first species of Paussus, which I have been long anxiously looking for: I took it in the daytime flying about fallen timber. Two days since I obtained a species of Malacomacrus, a Brazilian genus of Longicorns, described and figured by White in the 'British Museum Catalogue,'[75] and yesterday, while at breakfast, a magnificent black and yellow spotted Lamia flew into the verandah, and was caught in my hand. I have now 135 species of Bornean Longicorns, and I do not despair of getting 200 before I leave this place, which I mean to work thoroughly.

To give English entomologists some idea of the collecting here, I will give a sketch of one good day's work. Till breakfast I am occupied ticketing and noting the captures of the previous day, examining boxes for ants, putting out drying-boxes and setting the insects of any caught by lamp-light. About 10 o'clock I am ready to start. My equipment is, a rug-net,[76] large collecting-box hung by a strap over my shoulder, a pair of pliers for Hymenoptera [wasps and bees], two bottles with spirits, one large and wide-mouthed for average Coleoptera, &c., the other very small for minute and active insects, which are often lost by attempting to drop them into a large mouthed bottle. These bottles are carried in pockets in my hunting-shirt, and are attached by strings round my neck; the corks are each secured to the bottle by a short string. The morning is fine, and thus equipped I first walk to some dead trees close to the house frequented by Buprestidæ. As I approach I see the bright golden back of one, as he moves in

---

[75]  White, A. 1847–1855. *Catalogue of coleopterous insects in the collection of the British Museum*. London.

[76]  A misprint for bag net or rug net.

sideway jerks along a prostrate trunk,—I approach with caution, but before I can reach him, whizz!—he is off, and flies humming round my head. After one or two circuits he settles again in a place rendered impassable by sticks and bushes, and when he leaves it, it is to fly off to some remote spot in the jungle.

I then walk off into the swamp along the path of logs and tree-trunks, picking my way cautiously, now glancing right and left on the foliage, and then surveying carefully the surface of the smooth round log I am walking on. The first insect I catch is a pretty little long-necked Apoderus sitting partly upon a leaf: a few paces further, I come to a place where some Curculionidæ [weevils], of the genus Mecopus, are always seated on a dry sunshiny log. A sweep of my net captures one or two, and I go on, as I have already enough specimens of them. The beautiful Papilios, Evemon and Agamemnon, fly by me, but the footing is too uncertain to capture them, and at the same moment a small beetle flies across and settles on a leaf near me—I move cautiously but quickly on—see it is a pretty Glenea, and by a sharp stroke of the net capture it, for they are so active that the slightest hesitation is sure to lose the specimen. I now come to a bridge of logs across a little stream; this is another favourite station of the Buprestidæ, particularly of the elegant Belionota sumptuosa.

One of these is now on the bridge,—he rises as I approach,—flies with the rapidity of lightning around me, and settles on the handle of my net! I watch him with quiet admiration,—to attempt to catch him then is absurd; in a moment he is off again, and then settles within a yard of me; I strike with all my force, he rises at the same moment, and is now buzzing in my net, and in another instant is transferred in safety to my bottle: I wait a few minutes here in hopes that another may be heard or seen, and then go on; I pass some fallen trees, under which are always found some Curculionidæ, species of Alcides and Otops,—these I sweep carefully with my net and get two or three specimens, one new to me.

I now come to a large Boletus growing on a stump,—I push my net under it, two Thyreopteræ [ground beetles] run on to the top, I knock one with my hand into my net, while the other has instantly escaped into a crack in the stump and is safe for this day, but his time will come. In some distance now I walk on, looking out carefully for whatever may appear; for near half-a-mile I see not an insect worth capturing; then suddenly flies across the path a fine Longicorn, new to me, and settles on a trunk a few yards off. I survey the soft brown mud between us, look anxiously for some root to set my foot on, and then cautiously advance towards him: one more step and I have him, but alas! my foot slips off the root, down I go into the bog and the treasure escapes, perhaps a species I may never obtain again. Returning to the path, another hum salutes my ear, and the fine Cetonia, Macronota Diardi, settles on a leaf near me, and is immediately secured: a little further, a yellow-powdered Buprestis is caught in the same manner. Having reached the usual limits of my walk in this direction, I turn back and am soon rewarded by what appears a Colliuris sitting on a leaf, but which is discovered, on capturing it, to be of the equally acceptable Longicorn genus Sclethrus: a little further and a true Colliuris is caught.

These insects I have named, from their elegant form, lady-beetles, English names being necessary for the use of my boy Charley, who is now a rather expert collector. During the rest of the walk back, the principal insects I get are two velvety Elaters [click beetles] crawling on the logs, and two or three curious Heteromera in the same situation. Returning by the Chinamen's houses, I find, at an odoriferous puddle, the fine Papilio Iswara, which I capture, as well as a P. Evemon and P. Sarpedon. I then walk to the other clearing, where, among the fallen timber and branches, I get several small Buprestidæ; numbers of the handsome red Eurycephalus maxillosus are here constantly flying about and crawling on the timber. On one tree I find running about with ants, which they much resemble, the curious little short-elytra'd Longicorns, Hesthesis sp. Here also, I get two or three pretty species of Clytus and a Callichroma. Between whiles I have picked up a few flies, wasps and bugs, and have got tolerably filled bottles. Returning home, I find Charley has also had a fair day.

We empty our bottles into boiling water, and on pinning and setting our captures, find we have got between us 94 beetles, 51 different species, 23 of which are new to my collection: I have 5 new Longicorns, 2 new Buprestidæ, and 5 new Curculionidæ. I have been out five hours, and consider this a very good day's work. It will be seen that a far larger number of insects can be collected in a day in England, but perhaps hardly such a large proportion of species.

A.R. Wallace

---

## 11. To John Wallace, 20 April 1855

**Sarawak**

*April 20th. 1855*

My dear John

This is most sincerely to wish you joy of your marriage, and may you long live in unalloyed domestic happiness. I am sure Mary will make the best of wives: her good

temper, good sense, cheerfulness, and industry, are the very qualities to make all go smooth in her new home. I only wish, when my turn comes, as I hope it will, that I may draw as high a prize in the matrimonial lottery. Give her my kind love and a kiss for me, to be repaid when I visit you in California.

Fig. 7. John and Mary Wallace

May you go on prospering till then is the earnest wish
of your most affectionate brother, Alfred R. Wallace.

Mr. John Wallace

## 12. To George Robert Waterhouse, 8 May 1855
### Si Munjon Coal Works w. Sarawak
*May 8th. 1855*

My dear Sir

I *should* have written before to acknowledge the receipt of your letter containing your report on my Singapore Curculionidae [weevils],[77] and thank you for the trouble you took. I assure you it was most interesting & agreeable to me to find that some many of my insects were new, and has given me much encouragement to persevere in my search after the smaller species. You will already probably have heard through Mr Stevens that I have found another good locality here, which I continue to work very hard at. Curculionidae altogether do not bear quite so large a proportion to the other families as at Singapore, neither are the Anthribidae [fungus weevils] so numerous among them, but there is still a great number of fine things. Of *Cerastonema* I have two or three species all I think different from the beautiful one which you were so kind as to name after me.[78] *Micocerus Gazella* is very abundant, and there is another closely allied species which I hope may be new. The colouring of these insects is remarkably sober & untropical. Out of about 220 species of Rhyncophora which I have obtained here I have not one with metallic colours, nor indeed with any brilliant colours at all. I long to get into a country where I should find the beautiful *Pachyrhynchus* or something like it, perhaps I shall in *Celebes* or the Moluccas.

---

[77] Apparently, Waterhouse had examined some specimens sent back from Singapore. There is no evidence that a report was ever published, or that a copy was preserved.

[78] The genus '*Cerastonema*' was probably named in manuscript only by Waterhouse.

I turn from small beasts to large. I am now of course much interested in the "*Mias*" or *Orang Outan* & hope to determine the question satisfactorily of *one two or three* species. The three reputed species are found here & I think in two or three months I am sure to get adults of each for examination. I do not know whether you consider the three species as proved by the *sculls*, but it seems extraordinary that *Temminck* with the mass of materials at di his disposal should dispute it.[79]

At present my own opinion is that the small species *Simia Morio* (Owen) is undoubtedly distinct. I have seen the young of this & of a large species and

when fresh killed they could readily be distinguished though perhaps not by the stuffed skins. Of the existence of *two larger species* known by the single & double crested sculls I think there is still much doubt & this I hope to set at rest. The living animals are said to be distinguished by the presence or absence of the cheek callosities. Now the *callosities* do not seem always to accompany the *same form of scull*. Sir J. Brooke says the animal with cheek callosities has a double crested scull. Blyth[80] says the same, but Temminck

Fig. 8. George Robert Waterhouse

[79] The existence of three distinct species of orang-utan in Borneo was first suggested by James Brooke. 1841. Letter relating to the habits and points of distinction in the orangs of Borneo. *Proceedings of the Zoological Society of London* 1841 (13 July): 55–60. The division had not been recognized by Coenraad Jacob Temminck, director of the museum in Leiden, e.g. Temminck, Coenraad J. 1836. Sur le genre Singe – Simia (Linn.). *Monographies de mammalogie*. Leiden & Paris, vol. 2, pp. 113–40.

[80] Edward Blyth (1810–1873), curator of the museum of the Asiatic Society of Bengal in Calcutta (1841–1863). His views on orang-utan taxonomy were presented first in Blyth, E. 1853. Remarks on the different species of orang-utans.

(See Monographie de Mammalogie) has many skins with cheek callosities & the sculls belonging to them, but out of a large series only one double crested scull. Blyth says all the skeletons in Europe have the *single crested* scull, yet the animal has always been known by its cheek callosities as in the B.M. specimens. This would show that either the *crests* or the *callosities* or *both* are variable, & this I hope to be able to determine. I have got some sculls & skeletons from the Dyaks & have myself shot two small animals, but have not yet seen an adult of either species. I have however come to this place on purpose to get them, & as it is a good locality for insects I shall remain quietly for four or six months if necessary to settle this most interesting zoological question.

Any hints you can give me on the subject I shall be glad of & they will probably still find me here. I am getting also from my own observation & from the Dyaks, valuable information as to the habits of these most strange animals. I have had positive information that a species exists in the peninsula of Malacca very near Singapore & when I return there I shall endeavour to ascertain the fact.

I remain

Yours faithfully

Alfred R. Wallace.

G.R. Waterhouse Esq.

P.S. In the Lit. Gaz. You will see some account of my present locality.[81]

---

## 13. To Frances Sims, 25 June 1855

**Sadong River, Sarawak**

*June 25th. 1855*

---

*Journal of the Asiatic Society of Bengal* 22: 369–83, 10 pls. This was supplemented two years later in Blyth, E. 1855. Further remarks on the different species of Orang-utans. *Journal of the Asiatic Society of Bengal* 24: 518–28.
[81] Wallace, A.R. 1855. Borneo. *Literary Gazette* No. 2023: 683–4 [S22].

My dear Fanny

I will write my family letter this time to you, having just received yours of March 19th & my mothers of April 2nd. Do not write any more in the middle of the month, the only mail here now is on the 4th.[82] I was much pleased indeed to hear that you are getting on so well & that Thomas has at last got into the transparent pictures.

He has never told me yet whether there is anything new, or whether they are the same as he tried when I was with you & which I always told him would be sure to succeed if he would stick to them a little. You have not told me what you know I should be so much interested to hear, what he is to have for them. I trust he has not taken them too cheap. It is no good doing so. An order like that should make his fortune. If he has taken them at less than 5s each it is absurd. At that I suppose for a thousand he might make something handsome. In such an order he ought to do £100 for a month work.

I certainly think myself it would be unwise for you to go to another place in London unless Thomas has made up his mind to stay there eight or less years. The rent & expense of fitting up an establishment in a good situation would be so great as to use up all your savings up to this time & for such work as this, what better would you be off? He would be obliged to have more assistants if he meant to do a much larger business & would have much more anxiety & trouble. If indeed you could get a handsome sum for the present Glass House & apparatus then indeed you might venture but secure that first. If he has taken these clear pictures at a proper price at first & there are still as many portraits as Edward[83] can take & you still think of going abroad or at least getting away from London in a year or two, it certainly does not seem advisable to move.

If you do determine on risking the move to a better situation I hope you will live in the country at least 6-8 miles out of town near the Railway whose station

---

[82] Wallace refers to the mail leaving England on the 4th of every month.

[83] Edward Sims (1837–1906), a younger brother of Thomas. Probably first working as an apprentice, he opened his own studio in Westbourne Grove in 1857.

is nearest to your place of business,—where Edward could live. Then with 2 annual railway tickets you might be very comfortable, & all the long summer mornings Thomas could be *gardener* or *fowl keeper* & ruralize to his hearts content.

I am now obliged to keep pigs & fowls, or we should get nothing to eat. I have 3 pigs now & a China boy to attend to them, who also assists in skinning "orang-utans," which he & Charles are doing at this moment. I have also planted some onions & pumpkins which were above ground in three days and are growing vigorously. I have been practicing salting pork & find I can make excellent pickled pork here, which I thought was impossible, as everyone I have seen try has failed. It is because they leave it to servants who will not take the necessary trouble. I do it myself. I shall therefore always keep pigs for the future.

I find there will not be time for another box round the Cape, so must have a small parcel *overland*. I should much like my *lasts*[84] but nothing else, unless some canvas shoes are made. If the young man my mother & Mr. Stevens mentioned comes, he can bring them. I shall write to Mr. Stevens about the terms on which I can take him. I am, however, rather shy about it, having hitherto had no one to suit me. As you seem to know him, I suppose he comes to see you sometimes. Let me know what you think of him. Do not tell me merely that he is "a very nice young man." Of course he is. So is Charles a very nice boy, but I could not be troubled with another like him for any consideration whatever.

I have written to Mr. Stevens to let me know his character, as regards *neatness* & *perseverance* in doing anything he is set about. From you I should like to know whether he is quiet or boisterous, forward or shy. Talkative or silent, sensible or frivolous. Delicate or strong. Ask him whether he can live on rice & salt fish for a week on an occasion. Whether he can do without wine or beer, & sometimes without tea, coffee or sugar. Whether he can sleep on a board. Whether he likes the hottest weather in England. Whether he is too delicate to

---

[84]  A 'last' is a wooden form in the approximate shape of a human foot.

skin a stinking animal. Whether he can walk 20 miles a day. Whether he can work, for there is sometimes as hard work in collecting as in anything. Can he *draw* (not copy), can he speak French? Does he write a good hand, can he make anything, can he saw a piece of board straight? (Charles cannot & every bit of carpenter work I have to do myself.) Ask him to make you anything, a little card box, a wooden peg or bottle stopper, and see if he makes them neat, straight & square. Charles never does anything the one or the other.

Charles has now been with me more than a year & every day some such conversation as this ensues: "Charles, look at these butterflies that you set out yesterday." "Yes, sir." "Look at that one, is it set out evenly?" "No, sir." "Put it right then, & all the others that want it." In five minutes he brings me the box to look at. "Have you put them all right?" "Yes, sir."—"There's one with the wings uneven. There's another with the body on one side—Then another with the pin crooked. Put them all right this time." It most frequently happens that they have to go back a third time. Then all is right. If he puts up a bird, the head is on one side, there is a great lump of cotton on one side of the neck like a vein, the feet are twisted soles uppermost, or something else. In everything it is the same, what ought to be straight is always put crooked.

This after 12 months' constant practice & constant teaching! And not the slightest sign of improvement. I believe he never will improve. Day after day I have to look over everything he does & tell him of the same faults. Another with a similar incapacity would drive me mad. He never, too, by any chance, puts anything away after him. When done with, everything is thrown on the floor. Every other day an hour is lost looking for knife, scissors, pliers, hammer, pins, or something he has mislaid. Yet out of doors he does very well—he collects insects well, & if I could get a neat & orderly person in the house I would keep him almost entirely at out of door work and at skinning, which he does also well but cannot put into shape.

I must now tell you of the addition to my establishment in the form of an orphan baby, a curious little half nigger baby, which I have reared, now more than a month. I will tell you by and by how I came to get it, but in the meantime must relate my inventive skill as a nurse. The little innocent was unweaned and I had nothing proper to feed it with but rice water. I contrived a pap bottle with a large mouthed bottle, making two holes in the cork in one of which I inserted a quill so that the baby could suck. I fitted up a box for a cradle with a mat to lay upon which I had washed & changed every day. I fed it four times a day and washed it once and brushed its hair which it liked very much only crying when it was hungry or dirty. In about a week I fed it with a spoon & gave it the rice water a little more solid and always sweetened to make it nice.

I am afraid you would call it an ugly baby for it has a very dark skin and red hair, a very large mouth but very pretty little hands & feet. It has now cut its two lower front teeth & the uppers are coming. At first it would not sleep at night alone but cried very much, but I made a pillow of an old stocking, which it likes to hug and now sleeps very soundly, It has very strong lungs and sometimes screams tremendously so I hope it will live.

But I must now tell you how I came to take charge of it. Don't be alarmed, I was the cause of its mothers death. It happened as follows,—I was out shooting in the jungle and saw something up in a tree which of course I thought was a large monkey or orang utan, so I fired at it and down fell this little baby in its mothers arms. What she did up in a tree of course I can't imagine, but as she ran about in the branches very quickly I presume she was a "wild woman of the woods" so have preserved her skin & skeleton and am endeavouring to bring up her only daughter and hope some day to introduce her to fashionable society at the Zoological Gardens. When its mother fell mortally wounded the poor baby was plunged head over ears in a swamp about the consistence of pea soup and looked very pitiful. It clung to me very ~~tight~~ hard when I carried it home & having got its little hands unawares into my beard, it clutched so tight that I had great difficulty in making it leave go. Its mother poor creature had very long hair & while she was running about the tree like a mad woman, the poor little baby had to hold on ~~tight~~ fast to prevent itself from falling, which accounts for the remarkable strength of its little fingers & toes, which catch hold of everything with the firmness of a young vice.

About a week ago I bought a little monkey with a long tail, and as the baby was very lonely while we were out in the day time, I put the little monkey into the cradle to keep it warm. You will perhaps say (or my mother will) that this was not proper,—"how could I do such a thing",—but I assure you the baby likes it exceedingly, and they are excellent friends. When the monkey wants to run away by himself a little as he often does, baby clutches him by the tail & ears & drags him back, & if the monkey does succeed in escaping, screams violently till he is brought back again. Of course, baby cannot walk yet, but I let it crawl about on the floor a little to exercise its limbs, but it is the most wonderful baby I ever saw, and has such strength in its arms that it will catch hold of my trousers & hang underneath my leg for a quarter of an hour together without being the least tired, all the time trying to suck, thinking no doubt it has got hold of its poor dear mother. When it finds no milk is to be got, there comes another scream & I have to put it back in its cradle and give it "Toby" the little monkey to hug, which quiets it immediately.

From this short account you will see that my baby is *no common baby*, and I may safely say, what so many have said before with much less truth, "There never was such a baby as my baby"—and I am sure nobody ever had such a dear little duck of a darling of a little brown hairy baby before!

Madame Pfeiffer was at Sarawak about a year or two ago and lived in Rajah Brookes house while there.[85] Capt. Brooke says she was a very nice old lady something like the picture of *Mrs. Harris in "Punch"*.[86] The insects she got in Borneo were not very good, those from Celebes & the Moluccas were the rare ones for which *Mr Stevens* got so much money for her. I expect she will set up regular collector now, as it will pay all her expenses & enable her to travel where she likes. I have told Mr Stevens to recommend Madagascar to her.

---

[85] Ida Laura Pfeiffer (1797–1858), an Austrian widow traveller and author who visited many of the same islands in the Malay Archipelago before Wallace and who sold natural history specimens through Stevens.

[86] Mrs Harris was the fictitious friend of Mrs Gamp's in Charles Dickens's *Martin Chuzzlewit* (1843–1834). *Punch* used Gamp and Harris to lampoon the *Standard* and *Morning Herald*.

I have received the rings in the letter & am much obliged. Here I have no use for them, as so near Sarawak the Dyaks prefer money. When I take a trip further into the interior they will be useful, or in the Islands further Eastward. I also thank you beforehand for the things you have sent me by ship. I hope you secured the bacon well, filling up the Pot with bran & pasting a gumming paper round the edges. I am very glad I did not have anything to do with the Australian Expedition. Such tremendous delay & no well known man at the head of it. The recent news of Mr Strange's murder by the natives must be a rather disagreeable commencement.[87]

A gentleman has just come here to see the place who has recently come from England to join the mission. When he saw how we lived in open houses & open doors at night surrounded by Chineese & Dyaks he said "*People in England won't believe this.*" He said "I met a Dyak on the path with a long knife & I expected to have my head cut off." Whereas the idea of cutting off the heads of Europeans is never for a moment imagined by these poor people.

Let me hear about John & Mary when they arrive in California. I see by the Papers there have been some great failures there lately among others the House of Page Bacon & Co. through who he sent money, when I was in England. I trust he had nothing in their hands. Did you learn from him how he had invested his money, I hope safely but I suppose most in the *Water Company.*

Tell Mr Sims I have been dreadfully hard up for shoes lately & have been dissecting & patching till I am quite learned in the internal machinery of a shoe welts, ramps, quarters, uppers &c &c. I shall take care to keep a better stock by me for the future, as shoemaking without tools is hard work. I have never told you that my musical box is a never-ending delight to the Dyaks. They call it a bird and almost every day I am asked to show it them.

---

[87]  Frederick Strange (1826–1854) worked as a zoological collector in Australia. In October 1854, he sailed with the botanist Walter Hill and four other men to South Percy Island. All except Hill were murdered by six aboriginals, who were later hanged.

With my best love to my dear mother & remembrance to all friends, I must now remain,

Your affectionate Brother,

Alfred R. Wallace

Mrs. F. Sims.

## 14. To Frances Sims, 28 September and 17 October 1855

**Simunjon near Sarawak**

*Sept. 28th. 1855*

My dear Fanny

I received a short note from you & Thomas & from my mother by the May mail from which I learnt that you had taken suitable premises in Conduit St. at a Rent of £200 a year.[88] I imagined it was only a ground floor with place for glass house above & that you would live near Town. By the June mail I received no letters at all. By the July mail a short ~~mail~~ letter from my mother by which I hear that you are gone to live at Conduit St & have a whole house for your £200. By August mail which we expect here daily I hope to have some more particulars from you, & afterwards will finish this letter.

I am afraid you have been telling my mother not to write about your affairs or she would never have said so little as she has done in the last two letters. But why should you mind what she says to me? You know I don't judge hastily or harshly, & make full allowance for everything. It amuses her very much to write all she thinks at the time, & it amuses me here to read it, and as it is one of the

---

[88] Thomas and Fanny Sims moved to Conduit Street, London, in May 1855. An advertisement in the *Times* (23 July 1855) reads: 'MR. SIMS, photographic artist, begs to announce that he has REMOVED his PORTRAIT ROOMS from 44, Upper Albany-street to 7, Conduit-street, Regent-street. Photographs of every size and style, on glass, opal, enamel tablets, and paper. Also for the stereoscope, magic lantern, and oxy hydrogen microscope'.

chief pleasures she now enjoys to write at full to me, I should be sorry if you had given her any hint not to do so. However, perhaps you have not, only from what you said in one of your former letters I thought you might have done so.

Now as you live in Conduit Street of course my mother will have to leave Albany St. when the time is up and I should very much wish if you would try and arrange with her to get her settled comfortably so that she need not have to *move* any more which at her time of life must of course be very disagreeable & fatiguing. I think (& I recommended her in my last letter) that she should have a little cottage or half a small house somewhere near London, and near some of her friends (such as Miss Draper)[89] in some healthy & cheerful neighbourhood, where you could go down to spend Sunday with her now and then. I believe the House in Albany St will be given up in March next, when will be an excellent time for her to move. I think the plan would be much better & more comfortable for her & you, than for her to *lodge or board with any one*. It will give me a great deal of pleasure to hear that something of the kind has been agreed to & that she is comfortably settled.

I see in the Daily News 5 roomed cottages with garden near Peckham Rye advertised at £17 rent. Something of that sort would be just the thing. You could then spend from Saturday to Monday morning in the country.

In my mother's letter I received one of your new cards which is very *neat* & well arranged but to make it *handsome* wants a good wide margin. Get the next printed on a much larger card with a narrow blue border & a wide margin & it will do first rate. I hope to receive soon a full account of your opal & enamel pictures. You tell me nothing about your new things. I have not yet received a word of explanation about your transparent pictures except that they are much admired. I hope you have a fine set of specimens of all that you advertise. You should get Mr Vignolles[90]

---

[89] Miss Draper, a long-term friend of Wallace's mother, was the daughter of one of the executors of the will of Wallace's maternal grandfather, John Greenell (1745–1824). The matter was still unresolved in 1835 when Mary Ann Wallace wrote to Miss Draper on the subject (NHM WP1/3/4).

[90] Charles Blacker Vignoles (1793–1875), engineer, who in 1855 was elected as a Fellow of the Royal Society, and was a founder member of the Photographic Society of London.

or some other person to introduce some of your new things at the Royal Society's Soirées, or to bring some of "Athenaeum" or "Times" correspondent to see them. That's the way Mayall & Claudet[91] get so much talked about.

In a newspaper a month ago I received a paper photograph of a melancholy young lady, who should be much obliged to the person who folded the paper so cleverly as to *crease her exactly* across *both the eyes*, where there is now a faint hue of printing ink, of course vastly improving her personal appearance. Beside this the camera was below her face, which spoils a picture. I presume it is meant for Sarah Matthews but pray get her to smile or look cheerful when you take her next, & if you send me any more photographs of myself or any one else I should like to see a good clear sharp positive, if you can produce such a thing in Conduit Street. Everything you have sent me yet has been indistinct faint & washy.

Do you still burn your glass pictures, and are the opal pictures the unvarnished state or really done on opal? I hope you sent one of your new cards round to every one of your old customers, & I hope to hear of your taking from £5 to £10 a day now you are in a good situation. Do not fail to get a handsome plate at the corner of Regent St. as I recommended in my last; it would call in hundreds who, passing along Regent St. would otherwise know nothing of you. It would probably be worth more than £50 a year & perhaps produce £500 to you.

*Oct. 17th.*

The mail has just come in and again no letter from you or Thomas; three months you have not written! & you really don't deserve this letter, because you might write to me on Sunday, it would be a work of charity. However I have received a pretty long letter from my mother who gives some account of the unequalled magnificence of your rooms at Conduit St. quite a palace of

[91] John Jabez Edwin Paisley Mayall (1813–1901) and Antoine François Jean Claudet (1797–1867) were photographers experimenting with daguerreotypes and other new methods of photography in London in the 1850s.

Photography according to her account. But then I want to know how it pays, & you must have had nearly a month trial when she wrote, July 19th. To make up for your past delinquencies I shall look for a letter, 2 sheets crossed all over, entirely occupied with details of all your doings, new processes, prices, customers, what they are thought of, &c. &c &c &c &c &c &c

\* \* \* \* \*

*The box* has just arrived. The shoes fit well. The bacon I fear is not eatable. 8 months exposed to great heat, the fat melted & rusty, & all the small pieces utterly rusty throughout. Why did not you try lime, which succeeded with John? It would evidently be far cheaper to buy some ham or tongue put up by Fortnum & Mason on purpose for India & which will keep for years. If I want anything more I will send to them direct as it is a great pity to have such splendid bacon as it evidently was, spoiled. I am having a piece boiled for breakfast & will give you my account of it bye and bye. But anyhow I shall be obliged to eat it all quick.—

For the other things I am also much obliged & they will all be useful. I am in a hurry so cannot write more now so must enclose a note for my mother who I am glad to find intends as I suggest to live out of Town. I found in the box a glass picture for a white ground. Taken from a paper photograph I presume or from the glass. It is very pretty but the tone is blue & cold, though that hardly matters for buildings. Your circular (my mother encloses one) is very good & well worded. I hope your new pictures are a tall price.

In haste

Your affectionate Brother

Alfred R. Wallace.

*Mrs Sims.*

P.S. *After Breakfast.*

The bacon is eatable, *just*! But very high & very rich of a dark brown colour, which is a very nice relish, but I should have enjoyed a meal off it if fresh & I fear I shall be obliged to eat it too quickly to enjoy it. I suppose you bought mild bacon which could hardly keep 8 months good in a cupboard at home & this voyage is fully equal to 2 years in England.

I have written in a hurry & may have forgotten many things.

*AW.*

Via *Southampton.*[92]

Mrs. Sims

7 Conduit St.

Regent St.

*London*

# Sarawak

## 15. To Mary Ann Wallace, 25 December 1855

**Sarawak**

*Xmas Day, 1855*

My dear Mother

I have just received yours of Sept. 18th (why do you always write in the middle of the month when the post day is the 4th) and Fanny's of the 30th. You will see I am spending a second Christmas day with the Rajah and may probably not return to Singapore for near a month. I wrote hastily about 2 months ago acknowledging the receipt of the box with bacon &c. The Bacon turned out more eatable than I expected and was very useful as by frying a little it formed fat to cook fowl in.

Since then I have lived a month with the Dyaks & have been a journey about 60 miles into the interior. I have been very much pleased with the Dyaks. They are a very kind, simple & hospitable people, and I do not wonder at the great interest Sir J. Brooke takes in them. They are more communicative & lively

---

[92]  Postmark: SINGAPORE P.O. 22 NOV. [1855].

than the American Indians, and it is therefore more agreeable to live with them. In moral character they are far superior to either Malays or Chineese, for though head-taking has been a custom among them it is only as a trophy of war. In their own villages crimes are very rare. Ever since Sir J. has been here, more than 12 years, in a large population there has been but one case of murder in a Dyak tribe & that one was committed by a stranger who had been adopted into the tribe.

One wet day I got a piece of string to show them how to play *"scratch cradle"* & was quite astonished to find that they knew it better than I did & could make all sorts of new figures I had never seen.[93] They were also very clever at tricks with string on their fingers, which seemed to be a favorite amusement. Many of the distant tribes think the Rajah cannot be a man. They ask all sorts of curious questions about him, whether he is not as old as the mountains, whether he cannot bring the dead to life, and I have no doubt for many years after his death, he will be looked upon as a deity & expected to come back again.

I have now seen a good deal of Sir James & the more I see of him the more I admire him. With the highest talents for government he combines the greatest goodness of heart & gentleness of manner. At the same time he has such confidence & determination, that he has put down with the greatest ease some conspiracies of one or two Malay chiefs against him. It is a unique case in the history of the world, for a European gentleman to rule over two conflicting races of semi-savages with their own consent, without any means of coercion, & depending solely upon them for protection & support, and at the same time to introduce the benefits of civilisation & check all crime & semibarbarous practices. Under his government "running amuck" so frequent in all other Malay countries has never taken place, & with a population of 30,000 Malays, all of whom carry their *"creese"* & revenge an insult by a stab, murders do not occur more than once in 5 or 6 years.

---

[93]  Scratch cradle, or cat's cradle, is a game in which two people create a series of string figures.

The people are never taxed but with their own consent, & Sir J.'s private fortune has been spent in the government & improvement of the country, yet this is the man who has been accused of injuring other parties for his own private interests, & of wholesale murder & butchery to secure his government!

I am afraid my plants will arrive all dead, if so I shall send a few more overland. I have done very well in Borneo altogether but hope to do better in Celebes. I hope to hear good news of Conduit St. during the fine weather of the winter & spring when London will be full again. G.S. [George Silk] mentions a young lady "Miss Woodford"[94] whom you know in higher terms than I ever heard him speak of any young lady before, & then concludes by recommending her to me! if still single when I return. I am glad John is so busy & getting plenty of Dollars. When I go over I shall perhaps do a little surveying with him if it pays well. Of course it is nearer to go to California across the Pacific than by Europe & far less expensive.

I will write more fully when I return to Singapore. In haste now I remain,

Your ever affectionate Son<br>Alfred R Wallace

*Mrs. Wallace*
via *Southampton*[95]

Mrs. Wallace<br>7 Conduit St.<br>Regent St.<br>London

---

94    Miss Woodford was suggested by Silk as a possible future wife for Wallace. She has not been traced.

95    There are at least three postmarks on this page, and another partially printed, none decipherable. Another postmark reads: SINGAPORE P.O. *[illeg]*.

## 16. From Charles Darwin, [December 1855]

Fig. 9. Charles Darwin

— Skins Any domestic breed or race, of Poultry, Pigeons, Rabbits, Cats, & even dogs, if not too large, which has been bred for many generations in any little visited region, would be of great value, or even if recently imported from any unfrequented region. It w$^d$. be necessary to notice & select a characteristic specimen of *adult* animals of any breed.—In Poultry both cock & hen & especially the cock sh$^d$. be procured. The whole humerus & femur, & as much as possible of the cranium sh$^d$. be left on the skins.—

Each specimen of sh$^d$ be ticketed with native name, habitat & any procurable information. Specimens not bred for many generations in domestication are of no value.—

# Singapore

## 17. To Frances Sims, 20 February [1856]

**Singapore**

*Feb. 20th. 1855 [sic]*

My dear Fanny

I have now left Sarawak, where I began to feel quite at home, & may perhaps never return to it again, but I shall always look back with pleasure to my residence there & to my acquaintance with Sir James Brooke, who is a *gentleman* & a *nobleman* in the noblest sense of both words. I have just got yours & Thomas' letters of Nov. 30th. My foot got well at last after keeping me 3 months in the house. Camphor ointment did no good at all. Another time I shall use *caustic* which is the only thing in this country to make bad wounds heal.

I am sorry to hear you are not well. I hope you will not work too hard but take a days rest now and then and you should arrange to spend from Saturday afternoon to Monday morning in the country. Why did not my mother get a cottage & not take more rooms in London which I am sure she does not like so well as the country. I wish you would write me some more details of your business what are your highest prices, what is the most you take in a day, &c &c.

I suppose Thomas & you have been to see Fenton's Crimea Photographs[96]— the first great application of photography to *Life & History*. Do the transparent pictures for the gas microscope go on. No sooner do you seem to have got something new then I hear not a word more about it.

Charles has left me. He has staid with the Bishop of Sarawak,[97] who wants teachers & is going to try to educate him for one. I offered to take him on with me, paying him a fair price for all the insects &c. he collected, but he preferred to stay. I hardly know whether to be glad or sorry he has left. It saves me a great deal of trouble & annoyance & I feel it quite a relief to be without him. On the other hand, it is a considerable loss for me, as he had just begun to be valuable in collecting. I must now try & teach a China boy to collect & pin insects. My collections in Borneo have been very good, but some of them will, I fear, be injured by the long voyages of the ships. I have collected upwards of 25,000 insects, besides Birds, shells, quadrupeds & plants.

---

[96] Roger Fenton (1819–1869) took hundreds of photographs of the Crimean War in 1855. Soon after, he exhibited 312 prints in the gallery of publisher Thomas Agnew in London.

[97] Francis Thomas McDougall (1817–1886), first Bishop of Labuan and Sarawak from 1849 to 1868.

The day I arrived here a vessel sailed for Macassar, & I fear I shall not have another chance for two months unless I go a roundabout way, & perhaps not then. So I have hardly made up my mind what to do. The January mail is expected in daily so I may receive another letter from you before I send this. I have spoken to the Rajah about G. Silk. If matters go on well with the English Government there may be work for him here in a year or two. I shall write to him by this mail. This letter must do for all, as I have no time to write separately. I have sent a paper on Borneo & the Dyaks to the Geographical Society.[98] You will hear from G.S. [George Silk] when it is to be read & perhaps would like to go & hear it, as I have endeavoured to make it a little amusing & *readable* which the papers at the Geog. are not always.[99]

I think this war is a noble & a necessary one & it is only by its being thorough & complete that it can affect its purpose & secure the future peace of Europe. The warlike stores found accumulated at Sebastopol are alone a sufficient justification of the war. What were 4000 cannon for and other stores in proportion, if not to take Constantinople & get a footing in the Mediterranean, & ultimately to subjugate Europe? And why do such tremendous fortresses exist in every part of the frontiers of Russia, if not to render herself invulnerable from attacks which she has determined by her ambitious designs to bring upon herself? Russia is perpetually increasing her means of defense & of aggression; if she

---

[98] Wallace, A.R. 1857. Notes of a journey up the Sadong River, in North-West Borneo. *Proceedings of the Royal Geographical Society of London* 1 (6): 193–205. No letter by Wallace to the Royal Geographical Society is known from this period. However, in their fourteenth meeting of 23 June 1856, it 'was next announced that Mr. A.W. Wallace [*sic*], F.R.G.S., had returned to Singapore from his expedition to Borneo, and was preparing to visit Celebes, where he hoped to explore portions of that island hitherto unknown, as well as islands of the Molucca group. At the request of the Council, Mr. Wallace has been furnished, through the kindness of Lord Clarendon, with letters of introduction from the Governments of Holland and of Spain, to the authorities of their different colonies in the East' (*Proceedings of the Royal Geographical Society* 1: 97).

[99] A pencil annotation on the address page of the letter: 'A Paper sent by Alfred on Borneo & the Dyaks to the Geographical. We have heard nothing about it'.

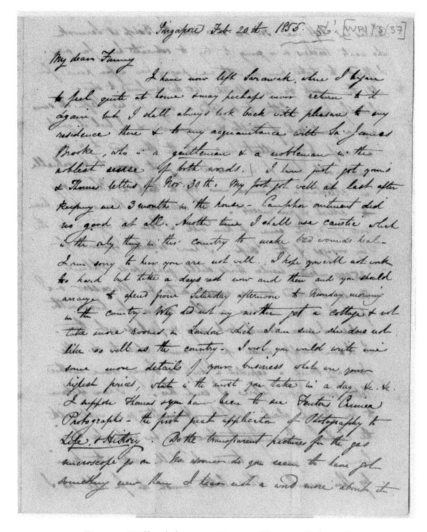

Fig. 10. Wallace's letter to Frances Sims, 20 Feb 1856

had continued unmolested a few years longer, it would have cost still greater sacrifices to subdue her. The war therefore is absolutely necessary as the *only means* of teaching Russia that Europe will not submit to the indefinite increase of her territory & power, & the constant menace of her thousands of cannon & millions of men. It is the only means of saving Europe, from a despotism as much worse than that of Napoleon as the Russian people are behind the French, in civilization.

Kind love to Mother, Thomas, Mr. & Mrs. Sims, Webster[100] & all friends.

In haste,

Your affectionate Brother

Alfred R. Wallace.

My dear Mother

I will write to you next time. Kind love from your affectionate Son

Alfred R. Wallace[101]

via *Southampton*[102]

Mrs. Sims

7 Conduit St.

Regent St.

*London*

## 18. To Thomas Sims, [c. 20 February 1856]

Dear Thomas

Your note about the engraving was very interesting. I hope you may make it succeed & keep it secret. You & Fanny talk of my coming back for a trifling

---

[100]  Possibly Mr. Webster, mentioned as a London builder in ML1: 9, the father of Mary Webster, married to Wallace's brother John.

[101]  The letter is annotated: 'A note for Fanny in this from Alfred'.

[102]  The address side shows four postmarks, one of which is partial and therefore illegible. The second postmark is illegible except for the date: 1856. The third postmark includes: SINGAPORE P.O. 22 FEB. 1856. The fourth postmark reads: INDIA PAID.

sore as if I was within an omnibus ride of Conduit St. I am now perfectly well, & only waiting to go Eastward. The far East is to me what the far West is to the Americans. They both meet in California, where I hope to arrive some day.

I quite enjoy being a few days at Singapore now. The scene is at once so familiar & strange. The half-naked chineese coolies, the neat shop keepers, the clean, fat, old, long-tailed merchants, all as busy & full of business as any Londoners. Then the handsome Klings, who always ask double what they take, & with whom it is most amusing to Bargain. The crowd of boatmen at the Ferry, a dozen begging & disputing for a farthing fare, the Americans, the Malays & the Portuguese make up a scene doubly interesting to me now that I know something about them & can talk to them in the general language of the place.

The streets of Singapore on a fine day are as crowded & busy as Tottenham Court road & from the variety of nations & occupations far more interesting. I

Fig. 11. Thomas Sims

am more convinced than ever that no one can appreciate a new country in a short visit. After 2 years in the country I only now begin to understand Singapore & to marvel at the life & bustle, the varied occupations, & strange population, ~~which~~ on a spot which so short a time ago was an uninhabited jungle. A volume may be written on Singapore without exhausting its singularities. "The Roving Englishman's" is the pen that should do it.[103]

---

[103] Pen-name of George Henry Augustus Sala (1828–1895), a travel writer known for his account of the Crimea for Dickens' *Household Words*.

Yours affectionately

Alfred R. Wallace.

T. *Sims* Esq.

---

## 19. To Samuel Stevens, 10 March 1856

### Singapore
*March 10th. 1856*

My dear Mr Stevens

I have received your letter of Jan. 6th. announcing the arrival in good order of the Insects by the "Cornubia".[104] At the same time I got the parcel of Books &c. which had been delayed a month *as usual* at Ceylon. The other shoes &c. do not send till I want something else. Do not send me more B.M. Catalogues, except new ones of *Coleoptera, Birds* & *Butterflies*. The moths have scarcely 50 Indian species in it. I send in the box a pair of broken spectacles. Get repaired at the *makers*, & get another pair exactly like to be sent in next parcel. Get one *Crawford's Malay Dictionary* 2 vols. (£2. I think) & send me the *2nd vol. only by post*, as the 1st is only a grammar & treatise which I do not want, but keep it for me.[105]

I have been here already a month, and most probably remain a month longer *doing nothing*. I can go nowhere for a short time, owing to the expense of travelling &c as it will be three months before I get more money from you, and after my

---

[104]  According to Baker (2001: 263), the '7th Consignment' of Sarawak specimens left Sadang on 21 July 1855 and Singapore by *Cornubia* for London in August 1855. The *Cornubia* departed Singapore on 20 August 1855 (*Straits Times*, 28 August 1855, p. 8) and arrived 25 December 1855 at Gravesend (*Times*, 26 December 1855, p. 7).

[105]  Crawfurd, John. 1852. *A grammar and dictionary of the Malay language, with a preliminary dissertation*. Vol. 2: *Malay and English, and English Malay dictionaries*. London: Smith, Elder and Co.

expenses here and the necessary outfit of clothes ammunition & other necessaries, which I *must now get*, I shall have little enough to pay my passage to Celebes & live for two months. Had I cash to spare I would make a collection of fish in spirits & go to some place on the main land, but I am afraid of any expenses, as I know the excessive inconvenience & misery of being without cash in a strange country. I ought to lay in a much larger stock of many things here than I can do, for if send for afterwards I shall get worse articles at double the price. The exchange too is now very bad for me, 4$^s$ 11¼$^d$ to the dollar the intrinsic value of which is only 4$^s$ 2$^d$.

Lacordaire's vol. 4 is as Bates says very good, send me the next vol. when out.[106]

You say the "Bangalore" coll$^{ns}$ are "*a very poor lot for Borneo.*"[107] But you must remember they were all collected in the wet season & before I had found a good locality, and if you turn to my letter when I sent them, you will see that I characterised them myself as a "*very miserable collection*" except the moths. I told you in my last that I would not trouble you any more about fixing prices, except in a few instances. Where there are a great number of duplicates you may dispose of some to the Paris dealers, but not till you have sold as many as possible to private parties and not till you have had the collection in your hands a good while, say 12 months. Never sell all the duplicates however. Keep back *10 or 12 always* for me.

The remarks of Count *[Mniszech]* are very amusing, but he is no entomologist to despise small insects. He errs as most people do in believing that the tropical insects are *generally* large & beautiful.[108] Mr. Payen's

[106] Théodore Lacordaire's *Histoire naturelle des insectes: genera des coléoptères.* Paris, 10 vols. 1854–1876.

[107] Wallace sent four consignments of specimens from Sarawak, which together left Singapore on the *Bangalore* on 25 April 1855 (*Straits Times* 1 May 1855, p. 7), see Baker (2001: 263).

[108] Wallace appears to have written Count Manyol. No person with a similar name can be identified and we suggest that Stevens may have written to him about the Count Georges Vandalia Mniszech (d. 1881), who had settled in Paris. Mniszech had a substantial collection of beetles, including many jewel beetles collected by Wallace, described in Deyrolle, Henri. 1864. Description des Buprestidés de la Malaisie, receuillis par M. Wallace. *Annales de la Societe Entomologique de Belgique* 8: i–vii, 1–272, pls. I–III.

coll.[109] which he mentions was the result of *many years'* collecting, not in Java only but over the *whole Archipelago*, & as it was no doubt principally obtained through natives, did not contain the smaller & more obscure species. The *show* therefore would be very great. But if at the end of 5 years collecting you pick out the finest things from my private coll[n] I have little doubt they will surpass it. Even from my Sing. & Bornean collections you may select a very fine series of large & handsome insects of which perhaps half will be single specimens. I believe that where any tropical country is as thoroughly worked as England is, the average size of the insects will be found to be no greater, neither will there be found a larger proportion of bright coloured ones. In a paper I enclose for *Newman*[110] I have said something on this point.

Before I left London the constant cry was "Do not neglect the small things." "The small things are what we want because they have never been collected in the Tropics."

By the *"Water Lily" left Singapore March* 5th. 1856[111]

2 casks with 5 orang skins in Arrack.[112]

1 Box, cont[g] 16 orang sculls & 2 skeletons

(six 6) 1 Box cont[g] a bundle of ferns & 6 boxes of insects & sundries

1 large case cont[g] Bird & Mammal skins, shells, reptiles,

1 box of insects & private papers

1 box cont[g] 2 boxes of insects. Books & papers & broken specs.

---

[109]  Antoine Payen (1792–1853) was a Belgian draughtsman and artist, interested in natural history, who donated a collection of Southeast Asian Coleoptera to the Leiden Museum in 1828.

[110]  Edward Newman (1801–1876), editor of *The Zoologist*. Wallace refers to his paper: Observations on the zoology of Borneo [dated 10 March 1856, Singapore]. *Zoologist* 14 (June 1856): 5113–17 [S25].

[111]  The schooner *Water Lily* left Singapore on 5 March 1856 (*Straits Times*, 11 March 1856, p. 8) and arrived in England on 21 June 1856. See *Notebook 1*, p. 116; and Baker (2001: 263).

[112]  A liquor made from fermented coconut sap. Wallace used it as a preservative spirit.

About the Orang skins I have written before, stating that I should wish if possible one purchaser for the whole series £250 or £300 (if in good condition), otherwise to be kept if they can be without any injury & with not much expense. *Consult Mess^rs Owen & Waterhouse.*[113]

The orang sculls & skeletons *are all private.*

Of the ferns, keep one complete set including the numbered specimens for me.

*The shells* dispose of as quickly as possible.

The Bird skins, in the *tin box* only, *60 in number,* are for sale, all others private

The mammals as before, sell, keeping back for me a series of the squirrels, & of all others which Mr. Waterhouse cannot name.

The Reptiles sell to any one, who will furnish me with a list of them. the whole! On those conditions they can go at a low price.

There are about 5000 insects for sale besides upwards of 2000 private. There are 2 boxes of Butterflies in papers, among which are a few good things. Of these you must select a series for me, if not already in my private coll. from Borneo. There are about 1500 more moths, which I worked very hard to get, staying alone up on the top of the mountain for a month or more. You must complete my series from them, & substitute better specimens where required, & you will then have left a fine series for Mr Saunders & for sale.[114] There are many fine new species among them & the small ones are better pinned or preserved than in my former coll^n.

---

[113] Richard Owen (1804–1892), comparative anatomist and superintendent of the natural history department of the British Museum.

[114] William Wilson Saunders (1809–1879), a wealthy insurance broker, insect collector, founding member and twice president of the Entomological Society of London. He contracted to buy Wallace's insects of all orders except Coleoptera and Lepidoptera. In the end Saunders probably purchased well over 10,000 of Wallace's specimens. See Baker, D.B. 1995. Pfeiffer, Wallace, Allen and Smith: the discovery of the Hymenoptera of the Malay Archipelago. *Archives of Natural History* 23 (2): 153–200.

*Value for Insurance*

I put the skeletons high because those sent before will be worthless without these sculls.

Should the first cask of skins have arrived & the skins be *spoilt*, reduce the amount above mentioned to £25.

A human scull in the large case is for J.B. Davis Esq., Shelton, Staffordshire, who will send for it.[115]

| | |
|---|---|
| 2 casks of Orang skins | £50 |
| 1 box of skeletons &c. | £100 |
| Insects (7000) | £75 |
| Mammal & Bird skins &c. | £25 |
| Total | £250 |

## 20. To Frances Sims, 21 April 1856

**Singapore**

*April 21st. 1856*

My dear Fanny[116]

I believe I wrote to you last mail, & have now little to say except that I am still a prisoner in Singapore & unable to get away to my land of Promise, *Macassar*, with whose celebrated oil you are doubtless acquainted.[117]

I have been spending 3 weeks with my old friend the French missionary,[118] going daily into the jungle, & fasting on Fridays on omelet & vegetables, a most

[115] Joseph Barnard Davis (1801–1881), a physician who amassed a large collection of human skulls and skeletons. Author of *Crania Britannica* (1856–1865).
[116] The letter is annotated: 'This note from Alfred I thought would amuse Algernon'. Algernon Wilson (1818–1884), Wallace's cousin who lived in Australia.
[117] Macassar oil was used for men's hair.
[118] Anatole Mauduit, who had been a missionary in Singapore since 1846.

wholesome custom which I think the Protestants were wrong to leave off. I have been reading Hucs travels in China in french, & talking with a french missionary just arrived from *Tonquin*.[119] I have thus obtained a great deal of information about these countries & about the extent of the Catholic missions in them, which is astonishing. How is it that they do their work so much more *thoroughly* than the Protestant missionaries?

In Cochin China, Tonquin, & China, where all Christian missionaries are obliged to live in secret & are subject to persecution, expulsion, & often death, yet every province, even those farthest in the interior of China, have their regular establishment of missionaries constantly kept up by fresh supplies who are taught the languages of the countries they are going to at Penang or Singapore. In China there are near a million Catholics, in Tonquin & Cochin China more than half a million!

One secret of their success is the cheapness of their establishments. A missionary is allowed about £30 a year, on which he lives, in whatever country he may be. This has two good effects. A large number of missionaries can be employed with limited funds, & the people of the countries in which they reside, seeing they live in poverty & with none of the luxuries of life, are convinced they are sincere. Most are frenchmen, & those I have seen or heard of, are well educated men, who give up their lives to the good of the people they live among.

No wonder they make converts, among the lower orders principally. For it must be a great comfort to these poor people to have a man among them to whom they can go in any trouble or distress, whose sole object is to comfort & ~~have~~ advise them; who visits them in sickness, who relieves them in want, & whom they see living in daily danger of persecution & death only for their benefit.

---

[119]  Évariste Régis Huc (1813–1860), a French missionary traveller and author of *L'Empire Chinois*, 2 vols. Paris: L'Imprimerie Impériale, 1854. There was a copy in the Singapore library.

You will think they have converted me but in point of doctrine I think Catholics & Protestants are equally wrong. As missionaries I think Catholics are best, & I would gladly see none others, rather than have as in New Zealand sects of native dissenters more rancorous against each other than in England. The unity of the Catholics is their strength, and an unmarried clergy can do as missionaries what married men can never undertake. I have written on this subject because I have nothing else to write about. Love to Thomas & Edward.

Believe me, Dear Fanny,

your ever affectionate Brother,<br>Alfred R. Wallace.

*Mrs. Sims*

---

## 21. To [Frances and Thomas Sims], [21 April 1856][120]

I shall be quite agreeable to take your Home News,[121] but you must arrange with the publishers to post it regularly to me instead of to you, or it may be too late for the mail. Enquire of Mr. Stevens for how long he has paid in advance & when that is finished he will pay the next half year to you, but mind & do not have any mistake so that it misses reaching me as usual.

I think you should discontinue your removal advertisement & substitute some more striking heading as,

Photographic Portraits on Enamel Opal Glass & Paper; Mr. Sims, 7 Conduit St. invites his friends & the public to examine his specimens in these new & beautiful styles, which he has at length succeeded in bringing to perfection.. &c &c &c…

---

[120]   Although this fragment is undated, it was most probably enclosed with the letter to Frances Sims of 21 April 1856. Thomas Sims had removed his photographic studio from 44 Albany Street to 7 Conduit Street in May 1855, and moved again in 1859 to 15 Newton Road, Paddington and later moved to Tunbridge Wells.

[121]   *Home news: a summary of European intelligence for India, China and the colonies*, published monthly in London from 1847.

I am sending home in a box with my insects some heavy cloth clothes (keeping a few for use in California) which you can take care of for me till my return. Last mail I only received one Athenaeum, but they generally come pretty regular. You did not acknowledge the receipt of the funny chinese pictures. I will get a few more now if I can. They will do to put in a scrap book. I have nothing more to tell you now but will write again when I have got a passage somewhere.

*ARW.*

## 22. To Henry Walter Bates, 30 April and 10 May 1856

**Singapore**

*April 30th. 1856*

My dear Bates,

Hearing from Mr Stevens that you had expressed a wish to hear from me, I will now do myself the pleasure of writing you a long letter, giving you an account of the Entomology of this part of the World, the details of which will be more interesting I am sure to you than to any other person to whom I could communicate them. I must first inform you that I have just received the "Zoologist" containing your letters up to September 14th. 1855 (Ega),[122] which have interested me exceedingly & have almost made me long to be again on the Amazon, even at the cost of leaving the unknown Spice Islands still unexplored.

I have been here since February waiting for a vessel to *Macassar* (Celebes), a country I look forward to exploring with the greatest anxiety & with expectations of vast treasures in the Insect world. Malacca, Sumatra, Java, & Borneo form but one zoological province, the *majority* of the species in all classes of

---

[122]   Bates, H.W. 1856. Letters from the Amazons dated between 12 September 1854 and 14 September 1855. *Zoologist* 14 (February): 5012–19.

animals being common to two or more of these countries. There is decidedly less difference between them than between Para & Santarem or Barra. I have therefore as yet only visited the best known portion of the Archipelago, and consider that I am now about to commence my *real work*.

I have spent 6 months in Malacca & Singapore, & 15 months in Borneo (Sarawak), and have therefore got a good idea of what this part of the Archipelago is like. Compared with the Amazon valley, the great & striking feature here is the excessive poverty of the Diurnal Lepidoptera [butterflies]. The glorious Heliconidae [heliconian butterflies] represented here by a dozen or twenty species of generally obscure-coloured Euploeas. The Nymphalidae [brush-footed butterflies] containing nothing comparable with the Epicalias, Callitheas, Catagrammas, Calleamura, Cybdelis &c. &c., neither in variety or number of species to make up for this want of brilliancy. *Termos clarissa* and a few species of Adolias, Limentis, and Charaxes are almost all. The Satyri-

Fig. 12. Henry Walter Bates

dae [brush-footed butterflies] have nothing to be placed by the side of the Haeteras of the Amazon. The glorious Erycinidae [metalmark butterflies] are represented by 5-6 species of Emesis and even the Lycenidae [gossamer-winged butterflies], though more numerous and containing some lovely species, do not certainly come up to the *Theclas* of Pará. Even the dull Hesperidae [skippers] are wanting here, for I do not think I have yet exceeded a dozen species of this family.

All this is very miserable, and is most discouraging to one who has wandered in the paths around Para & on the sands of the Amazon or Rio Negro. The only group in which we may consider the two countries as equal, are in the true Papilios (including Ornithop-

tera), though even in these I think you have more species. Including Ornithoptera
& Leptocircus I have yet got only 30 species (5 of which I believe are new). Among
them is the magnificent *Ornithoptera Brookiana*, perhaps the most elegant butterfly in the world.[123] To counterbalance this dearth of Butterflies there should be
an abundance of other orders, or you will think I have made a change for the
worse, and compared with Para only perhaps there is, though it is doubtful
whether at Ega you have not your Coleoptera quite as abundant as they are here.

But I will tell you what I have got & then you can decide the question, & let
me know *how* you decide it. You must remember it is just now 2 years since I
came into Singapore, & out of that time I have lost at least 6 months by *sickness*,
& voyages, besides 6 months of an unusually wet season at Sarawak. However
during the summer at Sarawak, I was very fortunate in finding a good locality
for Coleoptera, which I worked hard. At Singapore & Malacca I collected about
1000 species of Beetles, at Sarawak about 2000, but as at least half or perhaps
more of my Singapore species occurred also at Sarawak, I reckon my total
number of species may be from 2400 to 2500.

The most numerous group is (as I presume with you) the Rhyncophora [wee-
vils], of which I have at least 600 species, probably much more. The majority of
these are very small, and all are remarkably obscure in their colours, being in
this respect far inferior to our British series of species. There are, however, many
beautiful and interesting forms, especially among the Anthribidae, of one of
which (a new genus) I send a rough sketch.[124]

---

[123] Wallace was the first to describe the black and green butterfly from a specimen
obtained at the Rejang River by Captain John Brooke Johnson-Brooke (1823–1868),
in: Description of a new species of Ornithoptera: Ornithoptera brookiana Wallace.
*Proceedings of the Entomological Society of London* (April 1855): 104–5 [S16].

[124] The original of the sketch was probably an enclosure to the letter, no longer
present, but Bates copied it in a contemporary copy of the letter (see Fig. 13).
The figure is annotated, possibly by Bates: "Some specimens have the anten-
nae much longer" and: "Note. This species, on its arrival, has been named
'Cerastonema wallacei' by Waterhouse F.B." Wallace referred to this new
name in his letter to George Robert Waterhouse of 8 May 1855, but there is
no record that it was ever published.

The group next in point of numbers & to me highest in interest are the *"Longicornes"* [long-horned beetles]. Of these I got 50 species the first 10 days at Singapore, and when in a good locality seldom pass a day without getting a new one. Of Malacca & Singapore species I obtained about 160 species, at Sarawak 290, but only about 50 of the former occurred at Sarawak, so my Longicorns must in all now reach 400 species, or very near it. Of these the Lepturidae are most rare consisting of only 4 species, next the Prionidae 8 species, Cerambycidae about 80 species, and all the rest Lamiadae[125] which thus comprise more than ¾ of all the Longicorns.

One of the most interesting groups of these is the Genus *Glenea* consisting of graceful insects about ½ & ¾ inch long, most elegantly marked with spots, bands

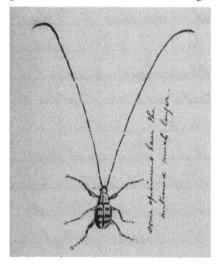

or lines of blue yellow or white. Of these, I have 45 species, fully ⅔ of which I expect will prove new as they are very active & are only found in shady forest paths, where I imagine scarcely any one has collected in Singapore or Borneo. In *Clytus* I am also rather rich possessing between 20 & 30 species. My largest species are two fine Prionidae 3¼ & 3⅓ inches long, a fine Batocera about the same size and a few fine species of Hameraticherus, Phryneta, Monohammus &c. In all, I have only about 30 species exceeding an inch

Fig. 13. Sketch in Wallace's letter to Bates, 1856

in length, the majority being from ½ to ¾ inch, while a considerable number are of 2 & 3 lines.

---

[125] Wallace writes Lamiadae instead of the correct name Lamiidae. The other names in this paragraph all refer to families and genera of long-horn beetles.

I see you say you must have near 500 species of Longicorns, but I do not know if this refers to *Ega* only, or to your whole amount of your South American collection.

The Geodephaga [carnivorous ground-beetles], always rare in the tropics, we must expect to be more so in a nearly level forest country so near the Equator, yet I have found more species than I anticipated. As near as I can reckon, I have a hundred. Of these, 24 are Cicindelidae, viz. 5 Cicindela, mostly small and obscure, 6 of the rare genus Therates, which are like small *Megacephalas*, & are found on leaves. Two or three of the curious apterous ant-like Tricondylas, and the rest belong to the elegant genus *Collyris*, resembling your Agias in form, but so active & ready to take flight, that it requires a most rapid & instantaneous stroke of the net to capture them.

Among the Carabidae [ground beetles] (besides the *Mormolyce*) the genus Catascopus is the finest, then *Orthogonius*, of which I have 6 species (5 alas! unique specimens!) and the rest are mostly small species of *Pericallus, Lebra, Dromius & Demetrias*, and 2 or 3 species of the curious genus Thyceopterus.[126]

Lamellicornes [beetles] are very scarce, about 140 species in all, of which 25 are Cetoniadae [rose chafer] (all very rare) and about the same number of Lucanidae [stag beetles].

Elaters are rather plentiful, but with few exceptions small & obscure. I have 140 species, one nearly 3 inches long and several of 1½ inch. The Buprestidae [metallic wood- boring beetles] are exceedingly beautiful, but the larger & finer species are very rare. Half my collection (110 species) are under 4 lines in length, though one, *Catoxantha bicolor*, is 2½ inches.

Two genera of Cleridae [checkered beetles] (Omadius and Stigmatium) are rather abundant, others rare; but I have gradually got together a nice collection of at least 50 species, which compared with the very few previously known from this part of the world is very satisfactory.

---

[126]    Frederick Bates added an annotation: 'W. must mean here "Thyreopterus" — There is no genus "Thyceoptera"— F. B.—'.

The groups already mentioned are those in which I take most interest & of which I have therefore most accurately separated the species.[127] The phytophaga form the bulk of the remainder of the collection and though pretty are generally very small. The Heteromera are next in number and contain hosts of closely allied metallic collared species, and a series of pretty ones near our *Melandrya*. Then come the *Malacodermus*, which are more numerous in individuals than in species. The *Brachyelytra* are very scarce & of *Paussus* I have not yet obtained a species. There are many curious Endomychidae & some Erotylidae, under decaying wood & fungi.

The individual abundance however of Coleoptera is not so great as the number of species would show. I can hardly collect on an average more than 50 beetles a day, in which there will be from 30 to 40 species. Often, in fact, 20 or 30 is as much as I can scrape together even when giving my whole attention to them, for unfortunately Butterflies are too scarce to distract it.

What us your greatest number of *species* of Coleoptera collected in a day;— mine is 70, of which 17 were Longicorns.

Of the other orders I have no very accurate account; the species however of all the orders united (except Lepidoptera) about equal the Coleoptera. I found one place only where I could get moths & obtained above a thousand species, mostly of small & average size. My total number of species of Insects I therefore reckon at about 6000, & of specimens collected above 30,000. From these data I think you will be able to form a pretty good judgment of the comparative entomological riches of the two countries, which I hope you will communicate to me *as soon as convenient*. The matter however will not be permanently settled, till I have visited Celebes, the Moluccas &c. which I hope to find as much superior to the Western Islands of the Archipelago as the Upper is to the Lower Amazon.

In other branches of Natural History I have as yet done little. The birds of Malacca & Borneo, though beautiful, are so common as not to be worth

---

[127]   The scientific names in this paragraph all refer to different families and genera of beetles.

collecting. With the Orangutan I was successful, shooting 15, & proving I think satisfactorily the disputed point of the existence of two species. The forests here are scarcely to be distinguished from those of Brazil but by the various species of *Calamus* (Rattan palm) and the presence of *Pandani* [palms], as well as by the rarity or absence of those Leguminous trees with finely divided foliage, which are so frequent in the Amazonian forests.

The people and their customs I hardly like so well as those of Brazil, but the comparatively new settlements of Singapore and Sarawak are not good specimens. Here provisions & labour are dear & travelling both tedious & expensive. Servants wages are high and the customs of the country do not permit you to live in the free & easy style of Brazil.

I keep a complete series of Coleoptera & Lepidoptera from each Island, so as to study the Geog. distribution. Mr. Saunders takes series of all the other orders of which he has undertaken to publish lists. I wish you could work some locality on the N. side of the river, so as to ascertain how far it separates distinct faunas. I trust you have your Obidos & Barra collections kept quite separate. The *birds* of the lower Amazon are very distinct on the two sides. I put a *locality ticket* to every one of my specimens.

Of my Longicorns Carabidae Buprestidae & Cetoniadae I have made figures and short descriptions sufficiently accurate to determine how many of the species I may obtain for the future, are identical with those of Malacca & Borneo. This enables me to send home my private collections when I leave each locality. In the future I intend keeping a daily collecting register of the number of species of each order & of each principal group of Coleoptera I capture.—

If you would do the same, you would I think find it interesting for reference & for comparisons between your various stations and between us at some future day. I once took 130 species of moths in an evening, at a lamp in a verandah. I hope you are keeping plenty of duplicates especially of your Longicorns as I hope some day to be able to make exchanges with you, and I have some idea of collecting all Longicorns, if I continue to find them abundant & can get duplicates enough to exchange for the species of other countries.

I must tell you that the fruits of the East are a delusion. Never have I seen a place where fruits are more scarce & poor than at Singapore. In Sarawak & at Malacca they are more abundant, but there is nothing to make up for the deficiency of oranges, which are here so sour & disagreeable that they would never eaten even in England. There are only two good fruits, the Mangosteen and the Durian. The first is very nice, but not deserving of the high place generally given to it. The durian is however a wonderful fruit, quite unique of its kind, & worth coming to the Eastern Archipelago to enjoy. It is totally unlike every other fruit. A thick glutinous, almond-flavoured custard is the only thing it can be compared to, but that it far surpasses. Both these however can hardly be had more than one or two months in the year, and in all towns & villages except far in the interior, are dear. The plantains even & bananas are poor, like the worst sorts of S. America.

If you should fall in with [Richard] Spruce, remember me most kindly to him & tell him I will write when I get into unknown ground.

*May 10th.*

The ship is at last in which I have been waiting for nearly 3 months, and in about a week I hope to be off for Macassar. The monsoon is however dead against us & we shall have to beat all the way, it will be probably a forty days passage. But then I hope to be rewarded. Celebes is quite as unknown as was the Upper Amazon before your researches & perhaps more so. In the B. M. [British Museum] catalogues ~~from~~ of Cetoniidae, Buprestidae, Longicorns, Papilionidae &c. there are no specimens from Celebes, & very few from the Moluccas, & the fine large insects which have long been known by the old naturalists & some of which have recently been obtained by Madame Pfeiffer give good promise of what a systematic search may produce.

Wishing you good success & hoping to have an interesting letter from you in due course,

I remain, Dear Bates,

Yours faithfully,

Alfred R Wallace

H. *Bates Esq.*

## 23. To Samuel Stevens, 12 May 1856

### Singapore
*May 12th. 1856*

My dear Mr Stevens

At last the expected vessel has arrived, having been sent to Bally & Lombock, & delayed by calms.[128] It will sail again in 4-5 days, so I shall not be able to get the next mail first. We have to go to Bally where we shall stay a few days, & shall try to pick up something. Our voyage against the monsoon will be about 45 days. You will see therefore that to get from Sarawak to Macassar ~~has~~ will have taken me (a month waiting at Sarawak, ½ month voyage, 3 months at Singapore, and 1½ month to Macassar) in all *6 months utterly lost & at great expense*. Such things people never reckon when estimating the profits of collectors.

However I will take all precautions that such a thing shall not happen again, & trust that the country I am now going to will repay me for it. I trust such instructions will arrive to Hamilton Gray & Co. by this mail as to enable them to forward me money without any more difficulty as 3 months expenses here & the necessary stores have made a great hole in the £100 they let me draw for.

I have made preparations for collecting extensively by engaging a good man to shoot & skin birds & animals, which I think in the countries I am now going to will pay me very well. When in Singapore before small shot was abundant, now I have had great difficulty in getting a bag at a high price. You had better therefore send me 2 bags of No. 10 shot and 2 of No. 6 with other shoes & 2000 percussion caps &c. in a month or two, in a strong packing case.

---

[128]   The Dutch bark *Kembang Djepoon* arrived in Singapore on 30 April (*Straits Times*, 6 May 1856, p. 8).

Ask Mr. Gould or Mr. Sclater if the remainder of *Bonapartes Conspectus*, cont$^g$ the Gallinaceae &c. is out, & if so send it me by post.[129] I enclose a letter for Bates which please forward him when you next write. The *Papilio Codrus*? at the sale, I see sold for more than £1, which makes me hope to do well with the fine & rare *Papilionidae* [Swallowtail butterflies] of Celebes & the Moluccas.

I have heard from Bowring[130] who tells me that the duplicates of his Java coll$^n$. are not numerous & that he will keep them all to exchange with me when I return to England. He comes home for his health by next mail so you will probably see him. Remember never dispose of *all the duplicates* of *any of my species* to the foreign dealers, always leave from 2 to 12 for me. I have been making a small collection of crustacea from the market. The small ones I can succeed with pretty well, but those of larger size will rot & fall to pieces notwithstanding all my care to dry them. Will the B.M. buy fish from here or from Celebes &c. There is a young man here who would make a collection. How did Madame Pfeiffer preserve hers. She must have had a good many to fetch £25.

I have just taken to Hamilton Gray's two boxes.[131] The larger one contains a lot of books principally of classics and Divinity about 170 volumes. There is a list of them at the top of the box. *Quaritsch* who lives in a court out of Leicester Square, a dealer in all dictionaries & prayer books will I dare say buy them, or another man in Holborn who deals almost entirely in school books.[132] You can

---

[129]  Charles Lucian Bonaparte's *Conspectus generum avium* appeared in Leiden in 2 volumes. Wallace's heavily annotated copy of volume one, dated 1850, is in the Library of the Linnean Society of London. The second volume appeared in three instalments in 1854, 1855, and 1856. There is no record that Wallace ever received these later parts.

[130]  John Charles Bowring (1821–1893), merchant in East Asia and amateur collector of insects. Troyer, J.R. 1982. John Charles Bowring (1821–1893): contributions of a merchant to natural history. *Archives of Natural History*, 10 (3): 515–29.

[131]  This was Wallace's '10th consignment' shipped on the *Dunedin*, leaving Singapore on 1 June 1856 (*Straits Times*, 3 June 1856, p. 8) and arriving in England probably late July 1856, see Baker (2001: 261).

[132]  Bernard Quaritch (1819–1899) was a bookseller and collector who had a shop on Leicester Square from 1842.

show the list to two or three such persons & accept the highest offer. I should think they will fetch from £10 to £20. ~~Once~~ If you can find no one to buy the whole, they must be sold by auction, & you can send the proceeds deducting expenses & your commission by an order on the Oriental Bank to *Mr. Geo. Rappa Jun'* (care of ~~W. Craal~~ Mr. W^m^ Kraal[133] at Hamilton Gray & Co. Singapore). He is the son of the collector who lived many years at Malacca, but has quarreled with his father & is very badly off. At the top of the box are two volumes of mine the *Cyclopedia of Nat. History*, which I find too bulky to carry about, so please keep it for me.

Insure for £20. Books for £10.

In the other case are 2 monkey skins, 38 crustacea, 70 small shells, and near 700 insects, for sale; with about 440 insects for private collection and 37 Birds all private; the monkey's scull is also private. There are also a few articles addressed to Mrs. Wallace and Mr Sims. I have put a complete set of the insects now collected in Singapore for my private coll^n but there are several among them of which I got *one specimen* only before, the least perfect of the two specimens may therefore now go to Mr Saunders if you will carefully select them. There are many pretty new things showing that Singapore is far from exhausted yet, & will furnish hosts of novelties to a resident collector.

The "Water Lily" will be near arriving I hope soon after you get this, & I trust the Orang skins will arrive in good condition. Remember the *spectacles* I mentioned ~~in my last~~ some time back to be repaired & 1 pair new. The indigenous wood is I think of a species of *Dracaena*. Lobb is still here waiting to go to Labuan.[134] He appears a first rate collector & has had great experience in the East.

---

[133] William Kraal is listed as a 'shipping clerk' at the firm of Hamilton, Gray & Co. in the list of principal inhabitants (*Singapore Almanack and Directory*, 1856, p. 59).

[134] Thomas Lobb (1817–1894), botanist on his third collecting trip in the East, travelling from Moulmein (Burma) and leaving for Labuan, Borneo. Lobb lived from collecting plants for the Veitch Nursery in England, well known for its abilities to raise exotic plants for British collectors and museums.

If the B.M. take any of the Crustacea, will you ask Mr. White[135] as a favour to furnish me with a list of them, as there are only about 15 or 20 species. The fish were most beautiful & varied & I longed to make a collection but I could not afford to lay out money on a cask of arrack. There are among the insects I think a few *diptera* [flies] which were not sent before, these must be added to the list. The few books I have returned, keep for me, I shall thus gradually reduce my luggage as I go on.

I think you will find that my male orangs are quite as large as that in B.M. & the large one in the last cask probably larger. At all events none of the specimens described by Temminck, Owen, or Blyth are larger than mine. But appearances are deceptive & every one believes they are larger than measurement shews them to be. Of course I wish you always to select for me first any *Lepidoptera* not already in my private collection to make the series for each locality complete. Also keep the localities in separate Boxes.

Drawing on you is a very expensive way of getting money here, as besides getting only a dollar for 4$^s$ 10¾$^d$, I am charged 2½ per cent commission by H.G. & Co. [Hamilton Gray & Co.] for negotiating the bill, which makes each dollar cost me 5$^s$/-.

I have not much more to say, but that I shall always like to hear what is wanted from the countries I am about to visit. I presume *good* specimens of even the common lories cockatoos &c will sell, and I expect to get a good many new & some very handsome birds. If I can reach the birds of Paradise Country (the Arroo Isles) I shall be able to prepare good specimens of those gorgeous birds, one of the greatest treats I can look forward to.

I shall have I think 4 or 5 months of fine weather at Macassar & can then if inclined go on to Amboyna & have a fine season there, as the seasons are reversed in all the islands E. of Makassar & Timor—by this going backwards & forward I hope to be able to escape the wet season for the next two years.

---

[135] Adam White (1817–1879), Assistant in the Zoology Department of the British Museum from 1835 to 1863, specializing in crustaceans and insects. See: Clark, P.F. & Presswell, B. 2001. Adam White: the crustacean years. *Raffles Bulletin of Zoology* 49 (1): 149–66.

I enclose a large letter to Bates which please forward to him by the first opportunity & I hope he will write to me in return.[136]

Lobb has been in Moulmein & Burmah which he says is the finest country he has been in for plants and he thinks all branches of Nat. Hist. but very expensive travelling & other expenses. He does not think much of the Moluccas for plants but I have great faith in them for insects & birds. I hope a new *Cheirotomus* or something equally fine *may* turn up, & I think I deserve one at least. My next letter will let you know what prospects I have & I trust to have good news to tell you.

Yours Sincerely,

Alfred R. Wallace

*Sam<sup>l</sup> Stevens* Esq.

---

[136]   Wallace refers to his letter to Henry Walter Bates, 30 April and 10 May 1856.

# BALI, LOMBOCK, AND CELEBES
## 23 May–16 December 1856

On 23 May 1856, Wallace and his assistants Ali and Manuel Fernandez, a Malaccan Portuguese bird stuffer, sailed from Singapore on the bark *Kembang Djepoon* ('Japanese blossom' in Malay). Twenty days later they anchored off Bileling on the north side of the small island of Bali. They had just two days to look around the country, where Wallace admired the rice cultivated on neat terraces cut into the hillsides. On the beach he collected a dark-coloured tiger beetle which 'could hardly be seen upon the grey volcanic sand'.[137] It was the second time he had found these beetles closely matching their background colour.

The journey continued and on 17 June 1856 the bark reached the bay of Ampanam on the west coast of Lombock. Here they would stay for about two and a half months until there was a ship to Macassar. This gave Wallace the opportunity to collect insects and birds near the coast and make some short excursions inland. While staying with a Malay man on the southwest coast Wallace first saw that Australian bird types extended as far west as Lombock. He was apparently informed that these birds were not found on Bali nearby or any further west. Asian bird types known on Java and Bali did not extend to Lombock. It was his first hint of the geographical separation of animal types, which has since become famous as the Wallace Line.

It was already well known that Asian animal types inhabited the western part of the Archipelago and that there were Australian types towards the east. Normally such very different faunas are separated by great barriers, such as an ocean or mountain range. Uniquely, in this case, there is no visible barrier of any kind.

---

[137] *Journal 1*, 1. Registered in *Notebook 2/3*, p. 18 as 'sea side, dark volcanic sand'.

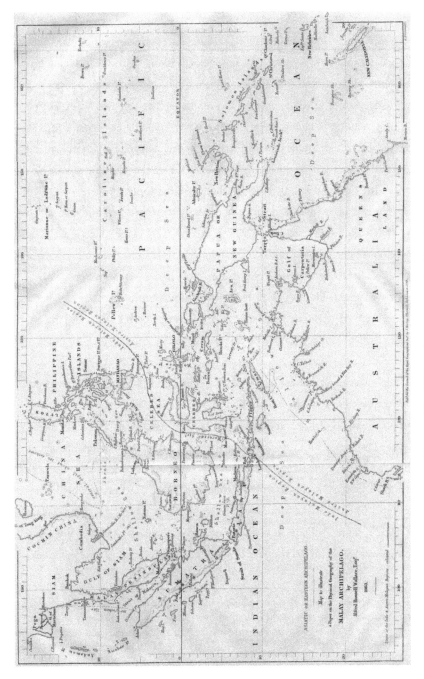

Fig. 14. The Wallace line, as conceived in 1863

Wallace eventually decided that the demarcation of animal types in the islands of the Archipelago was evidence of two ancient continents which had since subsided, leaving only fragmented landmasses as islands with this puzzling distribution. Wallace first mentioned his ideas on these biogeographical zones in a letter to Stevens of 21 August 1856. The same letter also contains Wallace's first surviving mention of contact with Darwin, when finally, after almost a year, he answered Darwin's request to assist him with specimens of exotic poultry and pigeons.[138] Some writers previously mistook this as the first correspondence between the two men and that Wallace had initiated contact.

At the end of August 1856, the schooner *Alma* arrived at Lombock which gave Wallace and Ali the opportunity to make their way to the island of Celebes, some 400 miles to the northeast. Fernandez had left Wallace's employ. Wallace and Ali reached Macassar on 2 September 1856. It was Wallace's first stay in a Dutch settlement. He brought a letter of introduction to a Dutch merchant, Willem Leendert Mesman, who lent Wallace a cottage on his farm not far from Macassar town. Wallace would use this as his base during his three-month stay in this part of the island.

Wallace was eager to hear how his views expressed in his Sarawak Law paper were received. From the letters in this section, we learn about the cautious consent of Sir James Brooke and the wholehearted agreement of Henry Bates. Brooke's letter of 4 July 1856 is one of the most important documents of this period to survive, because it provides unique contemporary evidence of Wallace's motives for publishing his famous Sarawak Law paper. Far from the bold statement of his belief in evolution that so many modern writers describe, Wallace wrote to Sir James in a now lost letter that the Sarawak Law paper was intended, in Sir James's words, 'to feel the pulse of scientific men in regard to this hypothesis'.

---

[138]  Darwin eventually discussed the colours of the skins of Lombok 'Penguin drakes', the bizarre upright Indian runner ducks, in *The variation of animals and plants under domestication* in 1868 (1: 280), although he noted they were sent by Sir James Brooke. This was the only time Darwin discussed any species from Lombok.

Wallace had also apparently stressed 'the bigotry & intolerance at which views or facts apparently adverse to received systems & doctrines are received'. This helps to explain why the paper was written in such ambiguous language and, as Wallace later put it, he left evolutionary descent 'to be inferred'.[139]

Wallace wrote a letter to Darwin on 10 October 1856, which unfortunately does not survive. However, from Darwin's detailed reply of 1 May 1857 we can reconstruct much of what Wallace must have said. He suggested that Darwin breed domestic varieties of poultry and pigeons himself to compare with the exotic breeds. He did not feel that sterility of hybrids would help much in settling the theory, which must have been a reply to an earlier remark by Darwin that is equally unknown today. Finally, Wallace mentioned that he did not believe that 'climatal conditions' determined the distribution of species. Lyell had argued that new species were created to suit the prevailing climate, but Wallace was convinced that the clue lay in the existence of antitypes, or ancestors, of a species. For him, the present geographical distribution of animals was explained by past geological connections over sunken land bridges. If climate was the determining factor then South America and South East Asia should have the same flora and fauna. As Wallace knew so well from personal experience, the two regions had fundamentally distinct fauna.

The November 1856 letter from Henry Bates contains lengthy details about his insect collecting in Brazil, which some readers will want to skip. Wallace, however, read it with great interest because it allowed him to compare numbers of insect species in different regions of the globe, and to judge if his collecting methods compared well with those of other collectors like Bates. Of more general interest are Bates' comments on Wallace's Sarawak Law paper. Bates highly approved of the conclusions, but added the cryptic line: 'The theory I quite assent to, &, you know, was conceived by me also'. It is difficult to be sure what exactly Bates meant by 'the theory'. Succession, the main point of Wallace's paper, was found in earlier

---

[139] Wallace, 1903. The dawn of a great discovery 'My relations with Darwin in reference to the theory of natural selection'. *Black and White* 25: 78.

works. Evolution of some kind was equally not new. Perhaps Bates referred to common genealogical descent, reading between the lines.[140]

# Lombock

## 24. From James Brooke, 4 July 1856

Sarawak

*4 July 1856*

My dear Wallace—

Being in a state of suspense and anxiety relative to some affairs with the British government I delayed answering the kind letter that you wrote from Singapore. You have left several friends in Sarawak and amongst them my poor self and I am only one.

I look back to the time you passed with us with much pleasure & the storm which has been raging for six years in my head has at long last blown over and has been succeeded by a gleam of sunshine. The British Government have withdrawn their Consul General and admitted some jurisdiction though Lord Clarendon[141] says they are *not prepared* to recognize me as an *independent sovereign*. To acknowledge the tribunals of a country is in fact to acknowledge its government but the government of England is pleased to run its head into a bush and hide this truth from its eyes which is glaring to others—it is this sort of shuffling action, to escape encouragement, which makes serious scrapes & begets difficulties by creating anomalies. Lord Clarendon as all the world says,

---

[140]   Wyhe, J. van. 2013. *Dispelling the darkness*, pp. 112–13.
[141]   George William Frederick Villiers, 4th Earl of Clarendon (1800–1870), Secretary of State for Foreign Affairs 1853–1859.

is a clever fellow, because Louis Napoleon has made a peace, but he is infirm of purpose and does not look a difficulty in the face.

I read your little brochure[142] with satisfaction, but I am somewhat misty on these subjects owing to being badly informed, but whether the successive development of species, i.e. the shading one into another so as to manifest a gradual onward series—or whether species be distinct and unconnected it seems to me that the metaphysical question is the same. The design (though we cannot understand or fit its parts) was complete from the beginning and it is an absurdity to maintain a supply of petty creations to mark the imperfection of the original work, for this is only to say, that there was no design and ergo no designer. The great machinery of God—or as some insist of nature—cannot be disturbed though dogs howl—women weep—or men pray—or to bring a mite or a mastodon into the world. The question however does not come to this as regards the mutation of species on the one hand and their complete separation on the other

for the advocate of unconnected species may argue that the germ has been ~~called~~ warmed into life when the proper & fostering conditions have arisen or arrived.

My great surprise is however at the bigotry & intolerance at which views or facts apparently adverse to received systems & doctrines are received—you say your little pamphlet is to feel the pulse of scientific men in regard to this hypothesis!— What ~~an utter outrage~~ an outrage of intolerance to need such caution?— It is this which makes me despair of

Fig. 15. Sir James Brooke

---

[142]    An offprint of Wallace, Sarawak Law, 1855.

advance—What harm can truth do us? What good can it not do us? and yet the enquiry is as beset with bristles as a porcupines back.—At any rate your paper was very short I may say that if you will prove this or any other hypothesis to my satisfaction I will receive it and I will never quarrel with any one who seeks for truth, because the search involves a scouting out my preconceptions & prejudices.—

You see how I have run away with my subject & my subject has ran away with me, or rather an outpouring of the spirit than the reason or reflection though the spirit suppressed has dwelt long & anxiously on these & similar topics. I must now tell you in brief that we progress famously and that the Borneo Company is an established fact which is supported by men of character & capital. The Bishop & St John[143] sailed yesterday to Labuan & Brunei—we have an immigration of 300 Chinese from Sambas. We have had Mr & Mrs Earl[144] staying with us—a pleasing addition to our society for a visit here is a favor conferred rather than bestowed. Your youngster Charles—now Martin is at Linga with Chambers[145]—they say he is not clever at books and when here he appeared damped & disheartened.—Let me hear from you and keep me informed of your changes of localities and believe me my dear Wallace.

Yours very sincerely,<br>Brooke.

*A. Wallace Esq.*

---

[143]  Spenser St John (1825–1910), Sir James Brooke's secretary and from 1851 Acting Commissioner and Consul General. St John wrote *Life in the forests* (1862), about his life in the east, and two biographies of Sir James.

[144]  George Windsor Earl (1813–1865), author of books on the Indian Archipelago and on the Papuans, who was married to Clara, née Siborne (b. 1830).

[145]  Rev. Walter Chambers (1824–1893) arrived in Sarawak in 1851 and went to work on the Linga River to build a church at Banting (McDougall, H. 1882. *Sketches of our life at Sarawak*, p. 41). Charles Allen assisted him for a while in 1856 after leaving Wallace's employ.

## 25. To Samuel Stevens, 21 August 1856

Ampanan

*August 21st. 1856*

My dear Mr. Stevens[146]

Another month has passed since I wrote to you & there is still no chance of a passage to Macassar. Having missed one opportunity by being away from the village, I am afraid to go out in the country any more, & here there are nothing but dusty roads, & paddy fields for miles around, producing no insects or birds worth collecting. It is really astonishing & will be almost incredible to many persons at home that a tropical country when cultivated, should produce so little for the collector.

The worst collecting ground in England would produce ten times as many species of beetles as can be found here, and even our common English butter-flies are finer & more numerous than those of Ampanam in the present dry season. A walk of several hours with my net will produce perhaps 2–3 species of Chrysomela [leaf beetles] & Coccinella, [ladybirds], & a Cicindela [tiger bee-tles], and two or three Hemiptera [bugs] & flies; and every day the same species will occur. In an uncultivated district which I have visited in the South part of the island, I did indeed find insects rather more numerous, but two months assiduous collecting have only produced me 80 species of Coleoptera![147] Why there is not a spot in England where the same number could not be obtained in a few days in spring. Butterflies were rather better, for I obtained 38 species, the majority however being Pieridae [white and sulphur butterflies]. Of the others, *Papilio Peranthus* [blue swallowtail] is the most beautiful.

The Birds have however interested me much more than the insects, as they are proportionally much more numerous, and throw great light on the laws of

---

[146] Samuel Stevens read an extract of this letter to the meeting of the Entomo-logical Society of London on 3 November 1856, published in the *Transactions of the Entomological Society of London* (new series) 4: 34.

[147] Wallace refers to Labuan Tring, which was a harbour in the bay of Lombock, cf. Heinrich Zollinger. 1851. *Journal of the Indian Archipelago and Eastern Asia* 5: 328.

geographical distribution of animals in the East. The Islands of Baly & Lombock, for instance, though of nearly the same size, of the same soil, aspect, elevation & climate, and within sight of each other, yet differ considerably in their productions, and, in fact, belong to two quite distinct zoological provinces, of which they form the extreme limits.

As an instance I may mention the cockatoos, a group of birds confined to Australia & the Moluccas, but quite unknown in Java, Borneo, Sumatra & Malacca. One species, however (Plyctolophus sulphureus [lesser sulphur-crested cockatoo]) is abundant in Lombock, but is unknown in Baly, the island of Lombock forming the extreme western limit of its range and that of the whole family. Many other species illustrate the same fact, and I am preparing a short account of them for publication.

My collection here consists of 68 species of Birds, about 20 of which are probably not found west of the island, being species either found in Timor & Sumbawa or ~~quite~~ hitherto undescribed. I have here for the first time met with many interesting birds, whose structure and habits it has been a great pleasure to study, such as the *Artamidae* [woodswallows] and the genera *Ptilotis, Tropidorhynchus, Plyctolophus* & *Megapodius.*

The islands of Baly & Lombock are inhabited by Malayan races, closely allied to the Javanese. Baly has several Rajahs, who are under the protection of the Dutch; Lombock has one Rajah, who governs the whole, & is quite independent.[148] These two islands are wonderfully cultivated, in fact, they are probably among the best cultivated in the world. I was perfectly astonished when on riding 30 miles into the interior I beheld the ~~whole~~ country cultivated like a garden, the whole being cut into terraces, & every patch surrounded by channels, so that [some words crossed out] any part can be flooded at pleasure.

Sometimes a hollow has the appearance of a vast amphitheatre, or a hill-side, a gigantic staircase, & hundreds of square miles of an undulated country have

---

[148]  The Rajah of Lombock from 1839 to 1870 was Gusti Ngurah Kketut Karang Asem.

been thus rendered capable of irrigation, to effect which almost every stream has been diverted from its channel & its waters distributed over the country. The soil is a fine volcanic mould of the richest description, and the result of such a mode of cultivation is an astonishing fertility. The ground is scarcely ever unoccupied; crops of tobacco, Indian corn, sugar cane, beans & cucumbers, alternate with the rice, & give at every season a green & smiling appearance to the island. It is only on the ~~hills~~ summits of the hills & on the tops of the undulations, where water cannot be brought, that the ground is left uncultivated, but in these places a short turf gives food to the cattle & horses, which are very abundant, & clumps of bamboos with forest & fruit trees have all the appearance of an extensive park, & are a pleasing contrast to the more regularly cultivated districts.

I have been informed by parties capable of forming a judgment that in the best cultivated parts of Java so much labour has not been expended on the soil, & even the industrious Chineese can show nothing to surpass it. More than half the Island of Lombock consists of rugged volcanic mountains, which are quite incapable of cultivation, yet it exports more than 20,000 tons of rice annually, besides great quantities of tobacco, coffee, cotton, & hides. Our manufacturers & Capitalists are on the look out for a new cotton producing district. Here is one to their hands. The islands of Baly, Lombock & Sumbawa can produce from ten to twenty thousand tons of cotton annually. It costs here *uncleaned* about 1½ cents a pound. The qualities are various, some, I believe, very good, so it can easily be calculated whether, after cleaning, it would pay.

So far will do for Newman[149] I think but first send it to my mother or sister as I have no time to write much to them. Tell Mr. Saunders that a Mr. Carter,[150] with whom I am living has applied for the situation of Lloyd's Agent here (100 sq. rigged vessels come here yearly) & I shall be obliged by his using his

[149]   Edward Newman was the editor of *The Zoologist*. Wallace was just starting a new page and was obviously aware that parts of his letters were going to be published in English magazines.

[150]   Joseph Carter, an English merchant on Lombock since 1855.

interest to get it for him. He is just the man, has been 18 years in the East as a commander of vessels & as a merchant, is well educated & would give any information that might be required of him. He would like to enter into the cotton *trade* if any one would communicate with him as to the expense of cleaning machine &c & he would send them samples of the cotton. Show Mr Saunders this, perhaps he knows some one interested in it.

I am now sending off to Singapore a case with my collections here.[151]

| Birds for sale about | 300 | Butterflies in papers | 150 | |
|---|---|---|---|---|
| Mammalia | 9 | Beetles | 250 | 465 |
| Land and freshwater shells | 100 | Miscellaneous | 65 | |
| a few sponges for Mr Bowerbank[152] | | | | |

My private collection of Birds are mostly in boxes separate & need not be opened: among the others, *all* which have a *red stripe* on the tickets, are *private*. Select a good series of the butterflies for me. The beetles I have separated.

Among the birds are a fine blue & white red billed kingfisher I think quite new, the pitta, bee-eater & orioles are also rare or new. Of the Malacca birds I have sent two or three of each for locality specimens. There are also some good specimens of the pigeons & generally the birds are very good specimens & I have taken great care to pack them. The domestic duck var. is for Mr. Darwin & he would perhaps also like the jungle cock, which is often domesticated here & is doubtless one of the originals for the domestic breeds of poultry.

You may insure the collection for £60 which I think is the lowest sum it will fetch as no collection of birds has ever been made here before & I am sure there are many new and rare things among them.

---

[151]  Wallace's collections made on Bali and Lombock were shipped from Singapore in the *City of Bristol* on 22 September 1856, see Baker (2001: 267).

[152]  James Scott Bowerbank (1797–1877), an English businessman interested in British fossils and sponges. He published *A monograph of the British Spongiadae*, 4 vols. (1864–1882).

Hoping soon to reach Macassar & to find good account from you there, I remain,

yours faithfully.

Alfred R. Wallace.

*Samuel Stevens* Esq.

# Macassar

## 26. To Samuel Stevens, 27 September 1856

**Macassar**

*Sept. 27th. 1856*

My dear Mr. Stevens

At length I am in Celebes! I have been here about three weeks & as yet have not done much, except explore the nakedness of the land. And it is indeed naked. I have never seen a more uninviting country than the neighbourhood of Macassar. For miles around there is nothing but flat land, which for half the year is covered with water & the other half is an expanse of baked mud (its present state), with scarcely an apology for vegetation. Scattered about it are numerous villages, which, from their being imbedded in fruit trees, have the appearance of woods & forests, but which, in fact, are little more productive to the insect collector than the paddy fields themselves. Insects in fact in all this district there are absolutely none.

I have got a bamboo house near one of these villages, about two miles from the town, which does very well for my head quarters.[153] To get into the country

---

[153] This 'little bamboo house' lent by Willem Mesman, at a place called Mama-jam, 'was situated about two miles away, on a small coffee plantation and farm, and about a mile beyond Mr. M.'s own country-house. It consisted of two rooms raised about seven feet above the ground, the lower part being partly open (and serving excellently to skin birds in) and partly used as a granary for rice. There was a kitchen and other outhouses' (MA1: 334).

is difficult, as it belongs to native princes, & there is no accommodation whatever for Europeans. There is, however, a patch or two of forest about 6-8 miles off & to it I have made several excursions, & got some birds & butterflies, but no beetles, which, at this season, seem altogether absent.

I cannot help comparing the facilities of the collector on the Amazon with the difficulties here. Whether at Parà, Santarem, Barra, Obidos or Ega, or any other town or village, you may always find good forest collecting-ground within a few minutes or half-an-hours walk of the place. You can live in the town & collect in the country round. In no place in the East that I have yet seen can this be done. Miles of cultivated ground absolutely barren for the naturalist extend round every town & village, & to get into the country with any amount of necessary luggage is most difficult & expensive. When there too the necessaries of life have all to be brought from the town, which renders living very ~~expensive~~ dear. The only way of moving is by means of porters or small carriages, the cost of which is about ten times that of boat hire, and in many cases you must expose yourself to the risk of life & property, being beyond the sphere of any civilized government.

However I hope soon to make arrangements for a small house near the forest I have spoken of, where I can stay a week at a time, & then bring home & store my collections at my house near Macassar. Already I see that I shall get a pretty good collection of birds. Raptorial birds are abundant (the first place I have seen them so in the Archipelago). I have already seven species, one or two of which I have no doubt are new. Of the forty species of birds I have already collected none are handsome, but several I think are new, among them a *Cinnyris* & a Pigeon. The rare parrot *Prioniturus platurus* [golden-mantled racquet-tailed parrot] is not uncommon here, though I have only obtained as yet only one specimen.

Among my few butterflies are two *Pieridae*, handsome & quite new, and two or three *Danaidae* [milkweed butterflies] which I do not remember to have seen. I have as yet got no *Papilios*, but do not despair of soon obtaining some fine ones. The place where I hope to do best is Bontyne about 60 miles from here. There is a road or path overland, but it would be very difficult to take all the luggage I require by that route, and by ~~water~~ sea, at the present time, owing to the wind being contrary, the voyage often takes from a fortnight to a month! In about January however the wind will be fair & the trip is then only 24 hours, when I shall probably go there, as I am informed there is plenty of forest, & the highest mountains in the island are close by.

The people here have some peculiar practices. 'Amok,' or, as we say, 'running a muck,' is common here. There was one last week: a debt of a few dollars was claimed of a man of one who could not pay it, so he murdered his creditor, & then, knowing he could be found out & punished, he 'run a muck,' killed four people, wounded four more, & died what the natives consider an honorable death! A friend here, seeing I had my mattress on the floor of a bamboo house, which is open beneath, told me it was very dangerous, as there were many bad people about, who might come at night & push their spears up through me from below, so he kindly lent me a sofa to sleep on, which however I never use, as it is too hot in this country.

On reading what I have written so far it may perhaps do for Newman [of the *Zoologist*]. I have received here your letters of May & June. I was astounded to hear that the Customhouse would make you pay for the orang spirit. My fishes from Para, Bates fishes &c never paid any duty, & I think that precedent should be urged, as had I known I should have used brine instead of spirit. Surely my character & yours ought to be sufficient, & I have no doubt an application to the Government with recommendations from some scientific men would be sufficient: however, you will of course have arranged it some how before this reaches you, & as I see by this mail that the "*Water Lily*"[154] has arrived, I hope by next mail to hear from you all about the whole lot, what [Richard] Owen thinks of the new species &c &c &c.

I told you long ago that you might dispose of all the Orang skins & skeletons as you could to the best advantage. Get Owen's advice, & represent the scarcity of good skins in spirits to enhance the price.

If you have not yet paid another £100 to Harvey Brand & Co.[155] on my account do so now, & send the order to pay me the amount to *Hamilton*

---

154 The schooner *Water Lily* arrived 21 June 1856 in England with the Sarawak consignment.

*Gray & Co.*, not to me here. I told you some time since not to send any parcel to me, yet as I shall probably want more arsenic, plaster of Paris & other things after leaving here which can all come together when anything is sent.

The place you mention as the locality for *Euchirus* [long-armed chafer beetle] appears to me to be a village in Amboyna, there is no island near of the name. M. Pfeiffers account of the mountains &c in Borneo can not be relied on; by the route she went there are no mountains more than 2-3 thousand feet high, & the men who were sent by Capt^n Brooke to accompany her into the Dutch territories said that she *never once* got from under the cover of her canoe all the voyage & could therefore see nothing whatever of the country or the people![156]

A friend of mine here who has done me many favours, wishes much for an English Rifle, so I promised to get you to order & send it. By this or the next mail you will receive a Bill of Exchange for £40, with a note from me giving the particulars of the Rifle which you can order of *Blisset 221 High Holborn*,[157] if he will do it for the price—if not *Loudon* will do it I have no doubt. The rifle complete is not to be more than £35 with ornaments &c as directed. The other five pounds is for packing case, carriage & for your time & expenses. It is to be sent overland directed to:

W. Mesman Esq. *Macassar*,[158]

care of A.L. Johnston & Co. *Singapore*,[159]

---

[155]   Harvey Brand & Co. was a mercantile house in London.

[156]   Wallace refers to Pfeiffer, I. 1855. *A lady's second journey round the world*, 2 vols. London.

[157]   John Blissett was a gun maker at 322 [*sic*] High Holborn, London from 1850 to the 1880s.

[158]   Willem Leendert Mesman (b. 1819) was a wealthy Dutch official who belonged to a powerful family based in Maros near Macassar, see Bosma, U. and Raben, R. 2008. *Being 'Dutch' in the Indies: a history of creolisation and empire, 1500–1920*. Singapore, p. 145.

[159]   A.L. Johnston & Co., merchants and agency house, established in 1820 by Alexander Laurie Johnston (d. 1850). It was the first European business in Singapore, and continued trading until 1892.

& to be finished as soon as possible. Any surplus cash to be spent on a riding whip & put in the case.

Tell Loudon to get my small double barrel gun, *80 bore*, which I ordered a month since, done as soon as possible, as persons here are waiting to see it, to send him further orders. Order two spare ramrods for it of *whalebone*, & get him to take off a discount for ready money.

I must now remain in haste,

yours very faithfully,<br>Alfred R. Wallace.

*Samuel Stevens* Esq.

## 27. From James Brooke, 5 November 1856

**Singapore**

*5 November 1856*

My dear Wallace,

You will be pleased to hear something of Sarawak, and the changes occurring there. Cruikshank[160] returned bringing with him a pretty and lady-like wife. Brooke and Charles Grant[161] have followed this good example and taken partners unto themselves, we expect them out about church next.

---

[160] Arthur Chichester Crookshank (1824–1891), cousin of Sir James and long-term resident in Sarawak. His return in October 1856 with his new wife Bertha was mentioned by Jacob, G.L. 1876. *The Raja of Sarawak*. London, 2: 331.

[161] Charles Thomas Constantine Grant (1831–1891), grandson of the seventh Earl of Elgin, who was Sir James's private secretary form 1848 to 1863. 'Brooke' was Captain John Brooke Johnson (1823–1868), nephew of the Rajah.

St. John I am sorry to say has been removed to Bruné. The government though yielding the question of jurisdiction and and [sic] not wishing to deny the right of the people of Sarawak to assert their independence would nigh acknowledge *my position*.—The concession has however been of great service and has rendered Sarawak particularly free & has restored me that peace of mind which I value more than these titles and honours. St. John however has been translocated and I am sorry for it.

The Borneo Company Limited[162] goes ahead nevertheless and I believe with a rational prospect of success. Coulson advances with his work—the rocks are out—and a steamer of size and speed, expected at about the end of this month. If they progress slowly and steadily, feeling their way, and making allowances for the undeveloped state of the country, I have good hopes for them, but if they expect each endeavour to command success, by the bold expenditure of capital, they will fail. My apprehension arises from neglect of proper means of [detail] to the end.—this is the stumbling block of civilized folk working blindfold in a rude country.

I came here for a change and can clearly benefit from it, not that I was ill though lower than I ought to have been considering how much I have to keep me in a high state of spirits.

I learn from Padday[163] at whose I am staying that you are now at Makassar after a sojourn at Bali. Before you leave this part of the world I hope to see you again at Sarawak which in its present stage of transition, may by that time present you with some new forms of animated existence—like our young ladies—as of ancient vegetable types, like myself. Have you read Baden Powells

---

[162] The Borneo Company Ltd., established in 1856, explored mining concessions at Simunjan and elsewhere in Sarawak. The local office was run by a Danish merchant, Ludvig Verner Helms, who had contacts with the trading-house of MacEwen & Co. in Singapore. See: Cox, H. and Metcalfe, S. 1998. The Borneo Company Limited: the origins of a nineteenth century networked multinational. *Asia Pacific Business Review* 4 (4): 53–69.

[163] Reginald Padday was a clerk at the firm of Hamilton Gray & Co. from 1855, and became a partner in 1856.

essays? Three in number viz. "Spirit of the Inductive Philosophy", "Unity or Plurality of worlds" and "Philosophy of Creation"[164] They are really philosophical and convey how a clergyman of the Church of England denote an apparent change in popular theology—accompanied as a matter of course with bigotry and /antifanaticism/.—The author utterly abandons the entire Mosaic Cosmogony—he alludes to "prudent philosophers" as suppressing truth for fear of consequences—adopts your view of the transmutation of species—as well as the Nebular theory not as proved, but as reasonable and religious probabilities, & in short takes large and satisfying views of Nature and of Nature's God.— Remember this little book—and make note of as Captain Cuttle says—it will interest you—I am sure.[165]

Charles alias Martin alias Allen[166] was miserable at the mission—the constraints were more than he could bear, which might have been foreseen had his previous life been considered before putting him into theological harness. He came to governmental employ though I had nothing for him to do, but I dare say he will get on in the employ of the Company who will work more acquainted with the language. The rest are all flourishing—I write you a desultory letter mainly to let you know that you continue in my thoughts and believe me

My dear Wallace, being sincerely yours.

J Brooke.

A Wallace Esq.

P.S. [Thomas] Lobb whom you have doubtless heard of is now in Sarawak.

---

[164] Powell, B. 1855. *Essays on the spirit of the inductive philosophy, the unity of worlds, and the philosophy of creation.* London.
[165] Captain Cuttle is a character from Dickens's *Dombey and son* (1848). Cuttle often repeated the phrase 'When found, make a note of'.
[166] Martin was the middle name of Wallace's assistant Charles Martin Allen (1839–1892).

## 28. From Henry Walter Bates, 19 and 23 November 1856

### Tunantins, Upper Amazon

*19 Nov<sup>r</sup> 1856*

Dear Wallace

I rec<sup>d</sup> your kind & very interesting letter—dated Singapore Apl 30 & May 10, on the 7<sup>th</sup> inst. as I was in the act of embarking on board the steamer at Ega for this village. You will be anxious to know what I have done in this new locality, but I will leave this to the end of the letter & proceed to give you some notes on what I have done since *Nov. 26, 1851*, the date of arrival at Santarem on my 2<sup>nd</sup> journey to the interior.—On the 17. Oct. last, at Ega, I made an enumeration of the species I had taken since Nov. 1851.—At that time I had my boxes empty, the whole of my private collection having been sent from Pará before I started from thence. Also at the end of 1854 I sent home my private collections of many groups (= Neuroptera—Bees & wasps & Ichneumon &c. Staphylini, smaller Carabides &c)—Explaining this I give you my statistics:—

| *"Lepidoptera"* | | *"Coleoptera"* | | [Coleoptera] 2357 | |
|---|---|---|---|---|---|
| Papilio— | 25 species | Cicindelides— | 38 species. | Lamellicornes— | 230 species |
| Heliconiae— | 44 | Carabides— | 272 | — viz. Copridae— | 112 |
| Pieridae— | 40 | Staphylini— | 120 | — Rutilae & Melolonthidae— | 75 |
| Nymphalidae— | 147 | Hydradepaha [and] | | — Dynastides— | 12 |
| Satyrides— | 85 | Phillydradae— | 30 | — Passali— | 12 |
| Erycinides— | 280 | Nitidules &c— | 60 | | |
| Theclae— | 140 | Hister, Scaphite— | 30 | Malacodermes— | 120 |
| Hesperides— | *192* | Poelaphi [and] | | Brenthi— | 30 |
| [Total]— | *953* | Seydmeni— | 25 | Coccinellides— | 30 |
| | | Cleri— | 84 | Scolytides &c— | 25 |
| Sphinges— | 8 | Heteromera— | 280 | Cyclica— | 620 |
| Castriae (?)— | 8 | Curculionidae— | 700 about | Erotyli— | *160* |
| Sesiae— | 16 | Longicornes— | 473 | [Total]—— | *3572* |
| Bombyelo— | 47 | — viz. Prioni— | 19 | | |
| Glaucophiles (?)— | 182 | — Cerambycidae— | 198 | | |
| Noctuides— | 148 | — Lamiae— | 251 | | |

| | | | | | | |
|---|---|---|---|---|---|---|
| Geometrides— | 76 | — Lepturae— | 5 | Coleoptera— | | 3572 |
| Micro-Leps:— | *150* about | | | Nocturnal Lepidoptera— | | 635 |
| | | Elaterides— | 150 | Diurnal Lepidoptera— | | 953 |
| [Total] | 635 | Buprestides— | *95 (⅔rds* under 3 lines long) | Other orders— | | *660* |
| | | [Total]—— | 2357 | Species of insects— | *5620* species | |
| Bees— | 43 species | = Fossores | 190 | = Homoptera— | | 40 |
| = Orthoptera | 15 | = Hemiptera | 260 | [Total] | 660 species | |
| Wasps | 12 | = Neuroptera | 40 | = Diptera | | 60 |

I have not separated the Santarem & Tapajos collections from the Ega one — but all the more important families are ticketed.—Yet I regret not having placed a "locality ticket" to *the whole* of the species, because I shall find some difficulty in fixing the geographical distribution, with certainty, in many cases.

Santarem—the *lower* Tapajos (to 100 miles up) & Villa Nova—are all very similar in their productions & form a strongly marked region in the Amazonian valley—extremely different from Pará-Cametá & from Ega. It is also different from the *upper* Tapajos (i.e. the river Cupari) which I explored well. There we find *many upper Amazon* species. I find a difficulty, however, at present, in forming a satisfactory generalization on this matter, because there are such things as a *difference in Station* (= soil, vegetation &c.) in regions, presenting, really, on the whole, but *one zoological character*. This difficult question I cannot discuss now but I hope to come to some deeply interesting results afterwards.

In the *upper* Tapajos about 8 or 10 of the *Ega* Nymphalidae [brush-footed butterflies] occur, but not numerously, the soil & forest is of the same character—great depth of vegetable mould and as a character of the vegetation, the cacao there flourishes.—In diurnal Lepidop. the genus *Calydna* (?) offers us excellent data.—The *Metiopolis* (?) of the 8 or 10 known sp. is Altar de Chão, 25 miles from Santarem.—In the *Cupari* only 1 species is found whilst at *Villa-Nova* nearly all the sp. occur.—At *Ega* I have only found *one individual* of the genus, a species, I think generally distributed & I believe we found it once or twice at Pará.—

What you say about the similarity of the species between Malacca & several of the Islands of the Archipelago—compared with the great difference we find at different points & ~~near~~ on opposite sides of the Amazon—suggests the hypothesis that *Central S. America* is a region of *elevation*—formerly consisting of Islands long isolated & containing separate Faunas—whilst the Eastern Archipelago is a region of depression with its opposite results—but I really do not know if the Archipelago is known by Geologists to be of this character. Without having the comparison of the European collections from different countries, I can form no satisfactory idea on these subjects—and this is a motive which will induce me to make a voyage to England before long.—

My list of species of Coleoptera [beetles] appears to be considerably larger than yours but it is the result of a much longer period collecting, & I think that in the same time you would get at least as large a number, so that, upon the whole, the 2 parts of the world are very similar as to *numbers of species*. As to *size & beauty*, you only will be able to say which is superior.

In *"Cicindelides"* [tiger beetles]—The genus *"Cicindela"* here—about 6 species—are all small & obscure, very inferior to *C. campestris* of England.—The *"Megacephalae"* are a splendid series—15 species—some *"Odontocheilae"* of the upper Amazon, are rich in metallic lustre. But, of this group, the *"Ctenostomae"* are the grandest. I have 9 species, 3 of which, ~~are~~ from Ega, are unique & probably new to science.

In *"Carabides"* [ground beetles] I think you would be astonished to see my coll., having found so few during your sojourn in this country. The *"Agras"*—15 or 20 sp.—of Ega are grand, some more than 1 inch long—others of most brilliant colors & curious forms.—All the *"Brachinides"* are curious, & very many fine things—numbers of *Calleidae*—*Lebiae* &c. I have also taken lately some things from Boleti, on high trees, w^ch appear to me to come near your *"Thyreopterae"*—They are very broad & flat, 4-6 lines long, yellow, orange &

black spotted. I have 2 very handsome sp. of those besides 3 "*Chelonodermae*" which are very similar.—& some *black* ones; besides, at least, a dozen other *metallic* ones w^ch are similar to "*Coptoderae*" but have not the *generic* characters of that genus. You will thus see that I have found plenty of *new forms* among the *Brachinidae*".—

In "*Scaritidae*" I have found a great number of species & some *quite new* genera.

In the "*Staphylini*" I have turned up a great no. of sp.—You will be surprised to see so many sp. of "*Cetoniadae*". They are chiefly "*Gymnetis*". I have found a dozen of a very fine *Allorhina* (?) reminding one of the Goliathi in the processes of head of ♂.

The list of "*Longicornes*" [long-horned beetles] is rich—I think I have 60 or 80 species from Carepi & Pará, in my coll. at London, not since found. They are not, as a collection, large or showy species; about 20 sp. pass 1½ inch—& ¾^th of the whole will be from 4 to 6 or 8 lines. Their general character is elegance of form & color.—You will see by part 2 of A. White's "*Longicornes*" (Brit: Mus: Cat:)[167] that there are *62 new sp.* of mine described—in the genera there treated upon, I feel sure that I have *30 more, new, unique* specimens.—Ega is very rich in *Longicornes*, & in *all* the Families of Coleoptera, it would be difficult indeed to find anywhere a spot so rich in Entomology: being especially rich in *Lepidoptera* as in *Coleoptera*. There are 18 sp. of *Papilio* within ½ a mile of the Town. New & fine things turned up daily after (altogether) 23 months of very close collecting & the last day I went out I found a *new Longicorn* ("*Clytus*").—

Since June last I register daily my captures.—In 1851 I did the same but on a defective place.—

Here is a copy of some of my day's work. I will firstly give you some of my *very best* days:—

---

[167]  White, A. 1847–1855. *Catalogue of coleopterous insects in the collection of the British Museum*. London.

= August 28/ 56 = Insects 206 (of which 114 minute Coleoptera)

Nov: sp:  1 Sphinx          N.S.    1 Coccinella (large)

1 large Bombyx                 1 Pinophilus (largest species yet taken)

1 Scarites

1 Macraspis                        1 Ophites (?)

2 Cyclopcephalae                   1 Hydrophilus

(very beautiful)                   1 Anthicus (horned sp.)

1 Cantharis (magnificent)          1 Scaritidae (curious thing)

2 Curculionides

1 Pselaphide (quite Nov: G:)

1 Clacigeride (new G:)

1 Opatride

1 Licinus ?         besides many new things in minute Pselaphidae.

all found on the beach after very high winds

1 Angocoris (?)

= *August 29* [1856] = Insects 120 (of which 35 minute Coleoptera)

N.S.    1 Prionide (quite a N.G. to me, near "Ctenoscelis")

1 *Mesomphaliae*

2 Curculionidae

1 Scarites

1 Philochlaenia (?)

1 Canthecona (?)     3 fine, large Hemipterae

1 Edeoside (?)

1 Mictide

(day's search on beach, after a high wind)

## Here are my last days in Ega:—

= *Oct. 24* = Insects 12, minute Col. 65

77 N.S. = 1 Erycininide, 1 Trogositide, 1 Hydroporus, 2 Curculionidae
warm & moist

= *Oct. 25* = Insects 26. *No Nov. Sp.* (warm & moist)

= *Oct. 26* = very heavy rain for 24 hours

= *Oct. 27* = Insects 30, minute Col. 10 = 40 N.S. 1 Brachide (grand)
sunny but very moist

= *Oct. 28* = Insects 20—N.S. 1 *Lamiidae* (very fine thing). sunny

= *Oct. 29* = Insects 25—N.S. 1 Lamiidae, 1 Brichide, 1 Bombyx, 1 Mantispa (?)
(hot—sunny between brief showers)

= *Oct. 30* = Insects 20 = N.S. 1 Longicorne (cloudy & warm)

= *Oct. 31* = (cloudy & warm, preparing for voyage)

*Resumé for October.*

Worked 25 days

Insects—Total specimens 835

New Diurnes 5

New Longicornes 7

New species 50 in all.

*Nov.* 2 = Insects 13 = N.S. 1 Longicorne, 1 Languria, 1 Tabanus

sunny

---

*Note.* = My no. of specimens is much less than yours. This results from my having already taken sufficient of all the species that are abundant in this Locality.

Now a few words about Tunantins.—I chose this place for a visit because it lies on the North bank of the river (Amazon), on the terra firma w^ch is continuous with the banks of the Inpurá up to the Andes, & is separated from Ega by the vast expanse of low, flooded lands forming the delta, of the Japurá, Juruá &c.—I thought also that 1 or 2 month search would decide whether the species change & become finer every 100 or 200 miles nearer the Andes, as our friends in London suppose.—I arrived here on the 11^th (Nov^r /56) & began to work on the 12^th, so I have had 8 days collecting.

I am sorry to find that insects of all kinds are very scarce, a fact w^ch I cannot explain as the grounds are most excellent—much varied—swamp—dry forest—ygapó [168]—clay soil—sandy soil—magnificent forest paths, in fact all that could be desired.—A good no. of the species w^ch first turned up were *new*, and when *I do find* a beetle in the woods, it is almost sure to be a new one: the conclusion is that it will require many month's stay to get a fair coll., but I cannot stay so long, for the immense no. of insect pests (clouds of "piumes" by day & mosquitos by night) added to *hunger* (for next to nothing is to be had to eat) are beyond my endurance.

In *diurnes* [true butterflies] I found at once 2 new "*Cybdeles*" very abundant & I have seen several of a 3^rd too nimble for me to capture as yet.—I have got 1 new "*Eubagis*", the largest of the genus—1 very distinct new "*Ithomia*"(?), & I

---

[168]  Ygapó is a term used for flooded lands in the Amazon region. See Bates, H.W. 1863. *The naturalist on the River Amazons.* London, vol. 1, p. 178.

see a new "*Timetes*" but cannot as yet capture it. I have also 2 new "*Theclae*", 2 "*Satyri*" & 2 "*Erycinides*".

In "*Longicorns*" but very few as yet, but 1 grand, new sp.—an "*Anisocerus*" 9 lines.—On *Nov. 12*ᵗʰ I took *49 specimens*—40 species—20 new. = *Nov. 13*: 70 = 39 species = 7 new. = Nov. 14: 39 = 35 species = 9 new. = Nov. 15: 70 = 39 species = 16 new. = Nov. 16: 39 = 38 species = 10 new. = Nov. 17: 14 = 14 species = 1 new. = Nov. 18: 40 = 35 species = 5 new. = Nov. 19: 17 = 17 species = 3 new.

I shall stay here about 6 weeks, at the end of which I shall be able to pronounce on the relations of the fauna with that of other districts.

I am now nearly at the end of my sheet without having touched upon any subject except Entomology although the *birds* & the *monkeys* of this upper river are interesting.—I am living, too, in the midst of a nation of Indians not yet reclaimed from the purely savage state.—They are the *Caishánas*, a very quiet harmless tribe.—There are about 300 of them, some of their houses are abᵗ a mile from the village, but the greater part live 2 days journey up the river Tunantins. They have no warlike weapons, & do not practice tatooing. They use, however, the Zarabatana & Urari poisons.—

There is another topic on wᶜʰ I must touch.—I recᵈ abᵗ 6 months ago a copy of your paper in the "*Annals*" on the "The Laws wʰ: have governed the introduction of new species-"[169] I was startled at first to see you already ripe for the enunciation of the theory. You can imagine with what interest I read & studied it, & I must say that it is perfectly well done. The idea is like ~~but~~ truth itself, so simple & obvious that those who read & understand it will be struck by its simplicity; & yet it is perfectly original. The reasoning is close & clear, & alth' so brief an essay, it is quite complete, embraces the whole difficulty, & anticipates & annihilates all objections. Few men will be in a condition to comprehend & appreciate the paper, but it will infallibly create for you a high & sound reputation. The theory I quite assent to, & you know, was conceived by me

---

[169]   Wallace, Sarawak Law, 1855.

also, but I profess that I could not have propounded it with so much force &
completeness.

Many details I could supply, in fact a great deal remains to be done to
illustrate & confirm the theory—a new method of investigating & propound-
ing Zoology & Botany inductively is necessitated, & new libraries will have to
be written—in part of this task I hope to be a laborer for many happy &
profitable years. What a noble subject w^d be that of a monograph of a group
of beings peculiar to one region but offering different species in each province
of it—tracing the laws which connect together the modifications of forms &
color with the *local* circumstances of a province or station—tracing as far as
possible the actual *affiliation* of the species. Two of such groups occur to me
at once, in Entomology, in *"Heliconiidae"* & *"Erotylidae"* of South America;
the latter I think more interesting than the former for one reason—the species
are more local, having feebler means of locomotion than the *"Heliconiidae"*.—

I accept your proposal of future exchange of specimens & had long time ago
thought of proposing it to you. In all the interesting families as *"Longicornes"* &
*"Carabides"* &c I shall continue to reserve specimens of all the sp. possible with
this view.

I have been badly furnished with copies of interesting papers & periodicals from England. I have not for a long time seen the "Zoologist" & know
not what rubbishing stuff they print from my hasty letters.—The papers
describing some of your things,—such as the *"Cleridae"* & the *"Diptera"*—
I do not see—I have got, however, the *proceedings of the Ent: Soc:* for 1855
complete, w^ch contain a great deal that was new to me & very useful to
know.

*Nov: 23^rd·* =

We expect the Steamer down every hour so I must conclude this letter. These
last 4 days have produced me *6 new Butterflies* & an Indian has brought me a
pair of a monkey (viz. *"Midis"* (?)) new to me.—The *best* of the Diurnes is a
*"Catagramma"*—very peculiar & rich in its colors, unlike anything fig^d by

Hewitson & as handsome as the finest of those figured in his work[170]—it somewhat approaches "*Lyca*"—There is yet another sp. flying about. The next finest thing is a "*Eurygona*" ~~also~~ likewise the handsomest of its genus, being *crimson*! with blk [black] border.—I have now 3 new "*Eurygonae*" here—*Coleoptera* are scarcer every day, only 1 new "*Longicorne*".

Yours very truly,

Henry Walter Bates

## 29. To Samuel Stevens, 1 December 1856

**Macassar**

*December 1st, 1856*

After this you will probably not receive another letter from me for six or seven months, so I must give you a full one now. I am busy packing up my collections here, but have been unfortunately caught by the rains before I have finished, and I fear my insects will suffer. The last four or five days have been blowing, rainy weather, like our February, barring the cold. In a bamboo house, full of pores and cracks and crannies, through which the damp air finds its way at pleasure, you may fancy it will not do to close up boxes of insects during such weather.

However, as the wet season has not regularly set in, we may expect a little sun and dry air soon, and then I am ready to pack and close everything. The neighbourhood of Macassar has much disappointed me. After great trouble I discovered a place I thought rather promising, and after more trouble got the use of a native house there, and went. I staid five weeks, and worked hard, though all the time ill (owing to bad water I think), and often, for days together, unable to do more than watch about the house for stray insects. Such a weakness and languor had seized me that often, on returning with some insects, I could hardly

---

[170] William Chapman Hewitson (1806–1878), a collector of butterflies who published *The genera of diurnal lepidoptera*. London, 3 vols. 1846–1852.

rise from my mattress, where I had thrown myself down, to set them out and put them away. However, now that I am back at my cottage near Macassar, with a few of the comforts of civilized life, I am nearly well, and will tell you what I have done.

My collections here consist of birds, shells and insects. In none of these, I am sorry to say, have I got anything very remarkable. The birds are pretty good as containing a good many rare and some new species; but I have been astonished at the want of variety compared with those of the Malayan Island and Peninsula. Whole families and genera are altogether absent, and there is nothing to supply their place. I have found no barbets, no Eurylaimi [broadbills], no Trogons, no Phyllornes [bulbuls]; but, what is still more extraordinary, the great and varied family of thrushes, the Ixodinae and the Timalias, seem almost entirely absent; the shrikes, too, have disappeared, and of flycatchers I have only seen one small species. To supply this vast void there is not a single new group, the result of which is that in about equal time and with greater exertions I have not been able to obtain more than half the number of species I got in Malacca. Indeed, were it not for the raptorial and aquatic birds I should not have one-third.

You hint that in Borneo I neglected Raptores. They are too good to neglect; but there were none. Here in two months I have got fifteen species, many more than all my collections of the two preceding years contain. Of these six are represented by single specimens only; but of the rest I send you thirty fine specimens, and they will, I doubt not, contain something new. Among my rare birds I may mention the two hornbills peculiar to Celebes (*Hydrocissa exarata*, Tem., *Buceros cassidix*, Tem.); the anomalous Scythrops Novæ-Hollandiae, *Lath.*; the handsome cuckoo, Phænicophaus callirhynchus; the Pica albicollis, *Vieill.*; and the remarkable Pastor corythaix, *Wag.*, which unites the characters of the starlings with the form and compressed crest of the Calyptomena and Rupicola.

My collection of land shells is at present very scanty; but then I have only been in one locality. It consists of five species of Helix, six of Bulimus, and one Cyclostoma. Of these I hope some will be new. There is a pale purplish Helix of the form of H. glutinosa, but in most specimens thickly speckled with blackish dots. Besides the common Bulimus citrinus, there are two closely allied species, one lightly marbled with brown near the base only, the other all over richly marked in a kind of zigzag pattern. Of both these I send a pretty good series. There are also, I think, three other small species, rather pretty, but very scarce. The Cyclostoma appears to be the same as the small, transparent, white one which was scarce at Sarawak.

Now for the insects, which are the most interesting to so many of my friends. They will, I fear, disappoint you, as they have, with a few exceptions, disappointed me. But you must remember the circumstances. Almost all the good insects have been collected during a five weeks' stay at a tolerable place in the interior, during which time, however, I was so unwell as not to make more than five visits to the forest, to be near which was the especial purpose for which I went there. It was also the very end of the dry season, which I have always found the worst time for insects. Notwithstanding these drawbacks, my collection presents some features of interest.

To proceed in order, the Coleoptera shall be first considered. The number of species yet obtained is only 254, some groups being rich, others very poor. My favourite Longicorns were so scarce as hardly to be worth looking for; yet among the few that fell in my way I have a new Agelasta, a fine Astathes, and a very curious insect with dilated thorax in the male, which will form, I think, a new genus, near Temnosternus, *White*. The Geodephaga are proportionately my richest group, as since the rains have commenced I have taken many curious small Carabidæ, among them three species of Casnoniæ. I am rich in Cicindela, having six species, but of Colliuris and Therates only one each. Cicindela Heros, *Fab*. (which I believe is rare) is my largest species. In Boisduval's 'Faune de l'Oceanie'[171] it is said to come from the isles of the Pacific. Therates flavilabris, *Fab.*, is also said to inhabit New Ireland, but it is found here, with the var.

---

[171] Boisduval, J.B.A.D. ed. 1832–1835. *Voyage de l'Astrolabe. Faune entomologique de l'Océanie*, 2 vols.

T. fasciata. The habitats given to insects in that work, indeed, from the French voyagers, appear so liable to error that little dependence can be placed upon them. They seem to have been trusted altogether to memory, or perhaps ticketed on the voyage home.

For example, to Scarabæus Atlas is this remark, "It is noted as from Vanikoro I., but M. D'Urville is certain that it was taken at Menado in Celebes;" again, to Tmesisternus septempunctatus, "If there is no mistake on the ticket, this species is from Amboina;" Lamia 8-maculata, "It is ticketed as coming from Vanikoro, but I believe it is rather from N. Guinea or Celebes;" and L. Hercules, "It is found in Amboina," while on the plate it is said to be from Celebes. Other examples of a similar kind are to be found; and they lead me to suppose that voyagers and amateur collectors seldom ticket their specimens *at the time of collection*, but trust to memory in a matter in which no memory can be trusted. Even after making a collection at two localities only, and of only a hundred species each, I would defy any one to ticket the whole correctly: how, then, must it be when dozens of places are visited in succession, and the species taken at each vary from perhaps a dozen to a thousand. But we must return to our collections.

In Lamellicornes I have been tolerably successful. I have found ten species of Cetoniadæ; a Tæniodera, common, I think; and the other nine all Protætias, a closely allied genus. All except one are small, and of that one (an inch long) I have only a pair, differing in colour, one black, one dark green, but, as they are marked with red exactly alike, I suppose them to be male and female. Among the smaller ones are some very pretty species and varieties, and of some of these I have a tolerable series. They are very local. All the best I got off one flowering shrub, which I visited daily for a week, when some heavy rains destroyed the remnant of the blossoms; and I never found another equally productive.

There is also a curious little brown thing, like a Trichius, which eats away roses and orange-blossoms. I have two Euchloras, which I think are rare, E. dichropus, Blanch., and a large one, very like E. viridis, but which seems to agree best with E. Dusumieri, Blanch. Besides these I have only a lot of obscure Melolonthidæ, Aphodii, &c., &c. I had quite forgot, however, among the Carabidæ, what will perhaps be considered my greatest prize, Catadromus tenebrioides; but it is very scarce. I have not found a single Lucanus; and the natives to whom I showed figures of them and other large insects, such as Scarabæus Atlas, denied their existence in the country; but no dependence is to be placed upon them, as they have not even a distinctive word for "beetle" in their language. In the other groups I have nothing particular, except a few pretty Rhyncophora and Phytophaga.

It is an ill wind, however, that blows nobody good; and the scarcity of Coleoptera will be highly satisfactory to some of my hymenopterist friends, since it led me to pay more attention to their favourite group than I should otherwise have been inclined to do. After the first showers fell, bees and wasps appeared in plenty, and I worked very hard at them. They are notoriously sunshine-loving animals; and for many an hour, when my health ought not to have permitted it, have I stood in the noonday sun, at some flowering shrub where they abounded, armed with net, pliers and bottle, intent upon their capture.

On the whole I have made, I consider, a very fine collection for such a very short time (less than two months). I have obtained in all 142 species, but of these 120 (about) are Aculeata, and, only about 12 being bees, the great majority are wasps, &c., of which many are very fine, large things, and the greater part seem to me different from any I took in Malacca or Borneo. I have also not neglected the small species, and I doubt not there will be a host of novelties.

The Diptera, Hemiptera, ants, &c., I have scarcely collected at all, but they promise well for another season. The Lepidoptera come last, and, though few in species, present a fair amount of novelty. On my very first visit to the forest I took three fine specimens of the magnificent Ornithoptera Remus, or a variety of it, for the female does not agree with Boisduval's very imperfect description of it. This made me think it common, but I have since never taken another, except an imperfect female. The common Ornithoptera here is a variety of Amphrisius, with the upper wings entirely black in both sexes.

Of Papilios I have three new species, one near P. Sarpedon, but the band narrow, dark green-blue on a velvety black ground, divided into rounded spots on the upper wings and linear ones on the lower. Of this I have a fine series. Another is close to P. Eurypilus, but quite distinct from all I have seen or that are described, by the abdomen above being pure white, which, with the white anal margin of the lower wings, and the white down which extends broadly over them, give the insect a most beautiful appearance when on the wing, and enabled me to pronounce it a new species the first time I saw it hovering over a muddy hole. It flies very strong, is rare and difficult to capture, and I secured very few specimens. The third is a rather obscurely marked species; near P. Helenus. I have only one specimen. Of Papilio Ascalaphus, Bois., I have taken the male and the female. P. Polyphontes is common, but I only obtained two or three good specimens.

Of Pieris and Euplœa I have several pretty things, and one or two good Nymphalidæ; but the best part of my collection, and what will perhaps please most, are the Lycænidæ, to which I have paid much attention. I have about 35 species out of 115 butterflies, and of half of these I have got the two sexes. With health, a better season and a better locality, I have no doubt a very fine collection of insects might be made in this part of Celebes, and these I hope to have next dry season, which I have arranged, if all goes well, to spend at Bontyne, situated at the South end of the Peninsula, and close to one of the highest mountains in Celebes.

I must now tell you where I am off to in the mean time. I am going another thousand miles eastward to the Arru Islands, which are within a hundred miles of the coast of New Guinea, and are the most eastern islands of the Archipelago. Many reasons have induced me to go so far now. I must go somewhere to escape the terrific rainy season here. I have all along looked to visiting Arru, as one of the great objects of my journey to the East; and almost all the trade with Arru is from Macassar. I have an opportunity of going in a proa, owned and commanded by a Dutchman (Java born),[172] who will take me and bring me back, and assist me in getting a house, &c., there; and he goes at the very time I want to leave.

I have also friends here with whom I can leave all the things I do not want to take with me. All these advantageous circumstances would probably never be combined again; and were I to refuse this opportunity I might never go to Arru at all, which, when you consider it is the nearest place to New Guinea where I can stay on shore and work in perfect safety, would be much to be regretted. What I shall get there it is impossible to say. Being a group of small islands, the immense diversity and richness of the productions of New Guinea will of course be wanting; yet I think I may expect some approach to the strange and beautiful natural productions of that unexplored country. Very few naturalists have visited Arru. One or two of the French discovery-ships have touched at it. M. Payen,[173] of Brussels, was there, but stayed probably only a few days; and I suppose not twenty specimens of its birds and insects are positively known. Here, then, I shall have tolerably new ground, and if I have health I shall work it well. I take three lads with me, two of whom can shoot and skin birds.

A. R. Wallace.

---

[172] Wallace called him Warzbergen (MA2: 197) or Herr von Abraham Warzebergen (in his unpublished *Journal 2*: 102). In the Dutch translation (*Insulinde*, 1871, vol. 2: 237), P.J. Veth states that the captain was of Dutch descent and called Van Waasbergen (alternatively spelled Waesbergen).

[173] Antoine Payen (1792–1853), a Belgian draughtsman and artist.

## 30. To John and Mary Wallace, 6 December 1856

**Macassar, Celebes**

*Dec. 6ᵗʰ. 1856*

My dear John and Mary,

Having luckily just heard of the birth of my Nephew, John Herbert Wallace,[174] allow me to congratulate you on the happy event and wish you and him many happy years. I say "luckily" because a few days more and I start for a place near New Guinea from which there is no return or communication for at least six months, would have delayed my knowledge of the great event.

I must also thank you for your kindness in giving me a new title, that of Uncle, which I trust to repay in kind, some of these fair days. When Johnny is three or four years old, and the others in proportion, you may expect to see me at Columbia. I presume the best time for me to arrive will be about February, so as to have the earliest spring fine weather, and be able to slope as soon as it gets too hot and dry.

Excuse me writing more now as am very busy preparing for my six months sojourn among the natives of Arru (a place you never heard of) and have lots of letter to write previous to so long an absence from communication with civilized humans, so Adieu, my dear Brother and Sister,

from your affectionate Brother

Alfred R. Wallace

To John and Mary Wallace.

P.S. Give Jonny a kiss each for me. A.W.

---

[174] John Herbert Wallace (2 August 1856–1914), the first son of John and Mary Wallace in Columbia, California.

## 31. To Frances Sims, 10 December 1856

**Macassar.**

*Dec<sup>r</sup> 10th. 1856*

My dear Fanny

I have received yours of Sept<sup>r</sup>, and my mother's of October, and as I am now going out of reach of letters for six months I must send you a few lines to let you know that I am well & in good spirits, though rather disappointed with the celebrated Makassar. My mother's letter has informed me that we are all now Uncles & Aunts, and I hope you enjoy the title. I hear your country residence is fixed on, & I hope it may lead to the reestablishment of your health. I am sorry however you have to pay so high a rent, because when you leave it, (& you cannot have any certainty of staying more than a year) my mother will have to move again, as even the *half rent* is too much for her in proportion to her income, & expenses of moving coming into *every year* must be a great burthen to her.

I believe my last letter to her & this will go together as the steamer is delayed & is probably broken down this month & I have told her all about where I am going so need not repeat it here. I have also written some details to Mr Stevens & G. Silk from either of whom you can no doubt get their letters.

The day after tomorrow I go on board the prow. We shall probably have 20 days passage, not very comfortable ones for me, but the prospect of the wonders & rarities to be obtained at its termination will keep up my spirits. For the last fortnight, since I came in from the country, I have been living here rather luxuriously, getting good rich cow's milk to my tea & coffee, very good bread & excellent dutch butter (3<sup>s.</sup> a lb.). The bread here is raised with *toddy*[175] just as it is fermenting, & it imparts a peculiar sweet taste to the bread which is very nice. At last, too, there is some fruit here. The mangoes have just come in & they are certainly magnificent. The flavour of is something between a peach & a melon, with the slightest possible flavour of turpentine, & very juicy. They

---

[175]  Palm wine.

say they are unwholesome & it is a good thing for me I am going away now. When I come back there will be not one to be had.

As I really know not what to write about to fill up this sheet, partly through having many preparations &c still to make & other letters to write, I will give you a list of my stores for my six months residence at Arru, so that you may see I am now taking pretty good care of myself.

| | |
|---|---|
| Sugar | 66 pounds |
| Tea | 7 " |
| Coffee (unroasted) | 27 " |
| Biscuits, 2 tins | about 20 " |
| Butter a small keg cont* | 11 " |
| Wine (madeira) | 18 bottles. |
| Pickles | 3 bottles |
| Fish sauce | 1 " |
| Oil for cooking | about 2 gallons |
| " for lamp | " 3 gallons |
| Vinegar (native) | 1 gallon |
| Soy (native) | ½ " |
| Jams & Jellies | *8 jars !!* |
| Black pepper | 1 pound. |
| Eggs | 100 … |
| Curry powder (home made) | 2 bottles[176] |

Rice all goes from here but the owner of the prow will supply me there. Fowls are scarce, but we shall eat Birds of Paradise. Fish are abundant, and there is a small *kangaroo* about as big as a hare which they say is very plentiful & excellent eating, so I expect we shall get on first rate in the culinary department. I also take with me a ~~bamboo~~ wicker chair & a small folding table, as I find sitting on the ground for months together dreadfully fatiguing & irksome for my long legs. I can neither write nor work with pleasure unless I have a chair & table.

---

[176]   A list headed 'Articles taken to Aru- & remarks on quantities' (*Notebook 2/3,* p.5) seems to be the original copy from which Wallace prepared this list. The separate list, however, gives additional items and adds his estimates of how long separate items would last his team.

I ought also to mention among my luxuries here, excellent *potatoes* grown on the Bontyne mountains & quite as good as you get in England. The sweet potatoes here however are so good that I like them as well or better & so take them by turns. Do not neglect to write by the *May mail*, as that will reach here just as I expect to return from Arru, & give me all the latest news. Kind regards to Thomas, Edw^d [Edward Sims], Mr & Mrs Sims & all friends. Has Eliza Roberts[177] got rid of her moustache yet? Tell her in private to use tweezers. A hair a day would exterminate it in a year or two without any one's perceiving. You have told me of nobody married or dead for a long time. Is the world stationary or are you too busy to notice it.

I remain, d^r Fanny,

> your ever affectionate Brother
> Alfred R. Wallace.

---

[177] Possibly "Miss Roberts, of Epsom, a cousin of my mother's" ML2:378. Thanks to Peter Raby.

# ARU AND AMBOYNA
## 17 December 1856–7 January 1858

One of Wallace's main aims was to reach the haunts of the extraordinary Birds of Paradise. This creature mainly lived in the unapproachable island of New Guinea, but some species were found on the equally remote Aru Islands located between New Guinea and Australia. His Macassar contact Mr Mesman told him about annual trading forays to the islands and introduced him to a half Dutch and half Javan man named Abraham van Waasbergen, the captain of a prau which was to leave soon for Aru. The traders stayed six months in Aru and returned to Macassar when the monsoon winds reversed. It was an excellent opportunity to reach Birds of Paradise economically.

Wallace boarded the prau together with 30 crew and 20 passengers, taking with him his faithful assistant Ali and two Macassar boys named Baderoon and Baso. They set out together with another prau on 18 December 1856 on their voyage of 1,000 miles. Two weeks later they reached the northern point of Ké Island, just short of their final destination. Here Wallace, for the first time, observed people of the Papuan race in their native region and was immediately convinced they were not related to Malays. Wallace collected birds and insects for six days until the prau was ready to set sail again. They reached the main Aru trading settlement of Dobbo on the small island of Wamar on 8 January 1857, Wallace's thirty-fourth birthday.

There were unfortunately no Birds of Paradise to be seen anywhere near the little trading settlement of Dobbo. Wallace had to make an excursion to the Wanumbai River on the largest of the Aru islands, where he spent close to two months. On 15 May, five days after his return to Dobbo, he wrote with excitement to Samuel Stevens that he had achieved his goal. He had obtained the Greater Bird of Paradise with its magnificent yellow, white, and maroon plumes.

Unlike the ragged specimens sometimes found for sale in the region, his were beautiful and glorious specimens that would make wonderful mounted displays. He also found some King Bird of Paradise: the smallest of all, but with a stunning crimson and white body adorned with emerald green tail wires that made it a 'perfect gem for beauty'. His notebooks record that Stevens would eventually sell the Aru collections for a handsome £360.

Wallace stayed on in Dobbo, often plagued with illness. Men came along the dusty streets selling food to the traders. Fish, rice, and sharks' fins were fine but Wallace was disgusted with sea cucumbers which he described as 'looking like sausages which have been rolled in mud and then thrown up the chimney'![178] He became annoyed with Baderoon who took his wages and quit, lost all this money gambling, and then even lost his freedom by becoming a debt slave. So only Wallace, Ali, and Baso left Aru on van Waasbergen's prau on 2 July 1857, landing back in Macassar ten days later. Wallace returned to his cottage at Mamajam, where he spent several weeks labelling and packing his Aru collections. He also wrote some important articles on the hitherto largely unknown natural history of the Aru Islands.[179]

Wallace now had time for a third period of concerted thinking and note taking on evolutionary theory. He shifted to an explicit focus on genealogical descent as the most plausible origin of new types of living things. He reflected in a short article written at this time 'on the theory of permanent and geographical varieties'.[180] In this piece Wallace represented himself as neutral on the question of evolution, although his arguments in fact pointed out how illogical it was to ascribe the origin of species to creation, but the origin of varieties to descent. It was, after all, admitted that species sometimes differed from each other less than some varieties did from their parent species. Such similar phenomena as slightly differing living forms, he argued, should not be attributed to fundamentally different kinds of causes.

---

[178] MA2: 201–2.
[179] Wallace, A.R. 1857. On the natural history of the Aru Islands. *Annals and Magazine of Natural History* (ser. 2) 20 (121, Supplement): 473–85 [S38].
[180] Wallace, A.R. 1858. Note on the theory of permanent and geographical varieties. *Zoologist* 16 (185–6): 5887–8 [S39].

In a letter written at Down on 1 May 1857, Darwin praised the Sarawak Law paper: 'I can plainly see that we have thought much alike & to a certain extent have come to similar conclusions. In regard to the Paper in Annals, I agree to the truth of almost every word of your paper'. Wallace was so flattered that he quoted this in a letter to Henry Bates (4 and 25 January 1858). Darwin also mentioned that 'This summer will make the 20th year (!) since I opened my first note-book, on the question how & in what way do species & varieties differ from each other.— I am now preparing my work for publication, but I find the subject so very large, that though I have written many chapters, I do not suppose I shall go to press for two years'. Most commentators now interpret this remark as Darwin's warning that Wallace should keep off his patch. This interpretation is based on the modern misreading of the Sarawak Law paper as an explicit avowal of belief in common descent. But Darwin had read the paper differently. He concluded that Wallace meant that new species were somehow 'created' based on similar predecessors—not genealogically descended from them. Hence Darwin had no idea that Wallace's private speculations were approaching his own so closely. In fact, Darwin was referring to his 20-year species project to more and more casual correspondents during this period since he was now deep into the process of writing up his big book on species theory.[181] So Darwin was certainly not warning Wallace away. Instead, Darwin was telling Wallace about his work-in-progress and giving an accurate estimate for its completion.

From 26 October to 8 November 1857 Wallace collected near the village of Maros to the north of Macassar marked by a striking formation of limestone hills. While there he wrote his first letter to Darwin to survive, if only partially. Wallace thanked Darwin for his letter, where he had expressed agreement on 'the order of succession of species' in the Sarawak paper. Wallace mentioned that he was disappointed that no notice had been taken of his paper and that he intended to write a book on the subject when he returned home. Just what this

---

[181]  Wyhe, J. van. 2007. Mind the gap: did Darwin avoid publishing his theory for many years? *Notes and Records of the Royal Society* 61: 177–205.

book would be about, apart from incorporating succession, was unclear. So the two men, although clearly in agreement to some extent and mutually sympathetic, were unaware of the details of one another's views. They certainly did not know enough of one another's ideas to feel like competitors. They were instead particularly sympathetic fellow naturalists.

At some unknown time Wallace collected further tiger beetles along the muddy banks of salt-water creeks in the vicinity of Macassar. Once again they were strikingly well camouflaged. The tiger beetles' 'colour so exactly agrees that it was perfectly invisible except for its shadow' (to F. Bates, 2 March 1858). This remarkable match puzzled Wallace.

On 19 November 1857, Wallace and Ali left Macassar aboard the monthly mail steamer. It followed the usual route, stopping at Coupang and Delli on Timor and then at the island of Banda, before it reached Amboyna on 30 November 1857. Wallace had a letter of introduction to a German physician, Otto Mohnike, the chief medical officer of the Dutch Moluccas and an accomplished entomologist. This gave Wallace some convivial company as the two naturalists compared specimens. After a few days, following Mohnike's advice, Wallace travelled to the middle of Amboyna for some serious collecting. It was here that Wallace found a large python in the rafters of his hut, an incident later dramatically depicted in his *Malay Archipelago*.

After some three weeks collecting, Wallace returned to the town on Christmas Eve. On 4 January 1858 he began a letter to Henry Bates, which in hindsight is surely one of the most important documents of the period, as it was written just a month before Wallace conceived of natural selection and wrote his famous Ternate essay. That theoretical breakthrough is the most intriguing part of Wallace's life, but unfortunately one for which we have almost no contemporary evidence to help us understand what happened in detail. Hence, modern commentators have offered many competing explanations to fill the gaps. In the letter, Wallace mentioned that Darwin's forthcoming book might save him the trouble of demonstrating 'that there is no difference in nature between the origin of species & varieties'. Wallace pointed out that fossil succession was linked to the geographical distribution of living species. History or ancestry was the key to understanding why certain parts of the Earth were inhabited by specific groups of organisms. A good example was the division of the Malay Archipelago between 'two distinct faunas rigidly circumscribed', referring, for the first time explicitly, to a 'Boundary line' between the two regions. That 'Boundary line' was dubbed 'Wallace's line' by T.H. Huxley in 1868.[182]

On the same day Wallace left Amboyna on the Dutch mail steamer heading north to Ternate, hoping to make it his headquarters and base to set out to New Guinea—the true homeland of the spectacular Birds of Paradise.

# Aru

## 32. To Samuel Stevens, 10 March and 15 May 1857
### Dobbo, Arru Islands
*March 10, 1857—*

Here I am, alive, well, and hard at work. I have been here just two months, and as I am going into the interior I leave this note to be sent by a vessel which returns to Macassar in April. The country is all forest, flat and lofty, very like the Amazonian forest. Insects, on the whole, are tolerably plentiful in specimens, but very scarce in species. There are, however, some fine things, and I am getting good series of several, including Ornithoptera sp., near Priamus, perhaps O. Poseidon, or close to it, a glorious thing but hard to get perfect; four or five other rare or new Papilios, but all are scarce; Cocytia d'Urvillei? rather scarce, a lovely creature; also Hestia d'Urvillei.

---

[182]  Huxley, T.H. 1868. On the classification and distribution of the Alectoromorphae and Heteromorphae. *Proceedings of the Zoological Society of London*, 1868: 313.

For six weeks I have almost daily seen Papilio Ulysses? [blue mountain swallowtail] or a new closely-allied species, but never a chance of him; he flies high and strong, only swooping down now and then, and off again to the tree-tops: fancy my agony and disgust; I fear I shall never get him. There is a fine Drusilla or Hyades abundant, with numerous varieties; but the Lycænidæ [gossamer-winged butterflies] and Erycinidæ [metalmark butterflies] are the gems; I only wish there were more of them; there are about half-a-dozen species equal to the very finest of the little Amazonians.

The Coleoptera are far too few in species to please me: in two months' hard work I can only muster fifty Longicornes, a number I reached in ten days in Singapore; but Lamellicornes are the most extraordinarily scarce; I have only nine species, and four of them single specimens; there are, however, two fine Lomaptera among them, I hope new. All other groups are the same; Geodephaga, scarcely a dozen species, and nothing remarkable; not one Cicindela; only one Tricondyla (T. aptera?) and one Thereres (T. labiata), with not a single Colliuris; two or three fine Buprestes, however, and some remarkable Curculionidæ, with the beautiful Tmesisternus mirabilis, make a pretty good show.

On my way here we stayed six days at Ké Island,[183] and I got there some very fine beetles, two fine Cetonias, and a Buprestis the most beautiful I have seen. Of the few insects I got there the greater part were different from any I have seen here, though the distance is only sixty miles, the mountains of Ké being visible from Arru in fine weather. This makes me think I shall get different things at every island in this part of the Archipelago. Arru is zoologically a part of New Guinea. Of the birds here half are New Guinea species; in the small island where we live many of the birds of Arru never come, such as the two species of the birds of Paradise, the black cassowary, &c. I am going now to the mainland, or great Island of Arru, in search of these birds, but have had the usual difficulty about men and boats.

---

[183]  Wallace was on the Ké or Kai Islands (Kai Besar) from 31 December 1856 to 6 January 1857.

I have learnt here all about New Guinea; parts are dangerous, parts not; and next year, if I live and have health, I am determined to go. I must go either to Banda or Ternate first, I have not yet decided which, and shall try and go to the large Island of Waigiou, at the north-east of New Guinea, where are found the Epimachus magnificus [scale-breasted paradise bird], three rare species of the Paradise birds, and the glorious Ornithoptera d'Urvilliana [a birdwing butterfly]? The weather here is very changeable; storm, wind and sunshine alternately. I think nine-tenths of the things I am getting will be new to the English collections; with which comfort for our entomological friends,

I remain yours sincerely,

Alfred R. Wallace.

*Postscript.*

**Dobbo**

*May 15*

I have returned from my visit to the interior, and the brig[184] is not gone yet; so I add a postscript. Rejoice with me, for I have found what I sought; one grand hope in my visit to Arru is realized: I have got the birds of Paradise (that announcement deserves a line of itself); one is the common species of commerce, the Paradisea apoda; all the native specimens I have seen are miserable, and cannot possibly be properly mounted; mine are magnificent. I have discovered their true attitude when displaying their plumes, which I believe is quite new information; they are then so beautiful and grand that, when mounted to represent it, they will make glorious specimens for show-cases, and I am sure will be in demand by stuffers. I shall describe them in a paper for the 'Annals.'[185]

---

[184] The ship was the 275 ton *Antilla*.

[185] Published as: Wallace, A.R. 1857. On the Great Bird of Paradise, Paradisea apoda, Linn.; 'Burong mati' (Dead Bird) of the Malays; 'Fanéhan' of the natives of Aru. *Annals and Magazine of Natural History* (ser. 2) 20 (120): 411–16 [S37].

The other species is the king bird (*Paradisea regia*, Linn.), the smallest of the paradisians, but a perfect gem for beauty; of this I doubt if any really fine specimens are known, for I think Lesson[186] only got them from the natives; I have a few specimens absolutely perfect. I have, besides, a number of rare and curious birds,—the great black cockatoo, racquet-tailed kingfisher, magnificent pigeons, &c.,—and a fair addition to my insects and shells. On the whole I am so much pleased with Arru that my plans are somewhat altered: on returning to Macassar I shall probably not stay more than two or three months, but get as soon as I can to Ternate, and then to the north coast of New Guinea, where all the remaining species of Paradise birds are found?

I believe I am the only Englishman who has ever shot and skinned (and ate) birds of Paradise, and the first European who has done so alive, and at his own risk and expense; and I deserve to reap the reward, if any reward is ever to be reaped by the exploring collector. I think there is good work for three years in N.E. Celebes, Gilolo Ceram, north coast of New Guinea, and intermediate islands, of all of which Ternate is near the centre, and it is certainly one of the least-explored districts in the world, and one which contains some of the finest birds and insects in the world. On the whole I have had much better health here than at Macassar, but I am now, and have been a whole month, confined to the house, owing to inflammation and sores on the legs, produced by hosts of insect bites. Confinement has brought on an attack of fever, which I am now getting over. My insect collecting has suffered dreadfully by this loss of time.—A. W.

---

[186] René Primevère Lesson (1794–1849), the first European naturalist to observe the mysterious Birds of Paradise alive during the voyage of *La Coquille* (1822–1825). Lesson, R.P. 1826–1830. *Voyage autour du monde, exécuté par ordre du roi sur la corvette de Sa Majesté, La Coquille, pendant les années 1822, 1823, 1824 et 1825. Zoologie.* Paris: Arthus Bertrand.

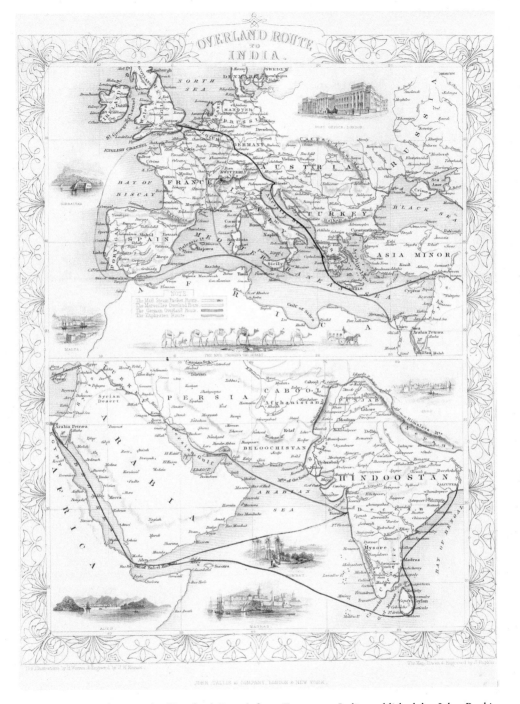

**PLATE. I.** Map showing the 'Overland Route' from Europe to India, published by John Rapkin (*The Illustrated Atlas*. London: John Tallis & Co., 1851).

Malay fishing village near Singapore. (John Cameron, *Our Tropical Possessions in Malayan India*, 1865, p. 129)

PLATE. 3.  The whites and sulphurs belonging to the butterfly family Pieridae were given much attention by Wallace. In 1867 he published a paper on them in the *Transactions of the Entomological Society of London*, where he described 45 new species, partly collected by himself. Five of these are shown here (plate 6): *Terias celebensis* (Macassar, Menado, Sulla Islands), *Pieris tamar* (Bali), *Pieris narses* (Australia), *Thyca pandemia* (Borneo), and *Thyca parthenope* (Singapore and Borneo).

PLATE. 4.  Orang outangs as Wallace would have known them before he set out for the Malay Archipelago. This plate was published in a French illustrated encyclopaedia of natural history, edited by Félix E. Guérin-Méneville (*Dictionnaire pittoresque d'histoire naturelle*, 1838, pl. 428).

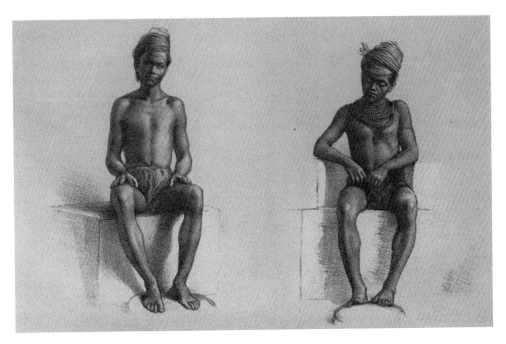

PLATE. 5. Two land dyaks, the original inhabitants of Borneo. (Spenser St. John, *Life in the forests of the Far East*, 1862, p. 135).

PLATE. 6. Three species of longhorn beetle (Coleoptera, Cerambycidae) collected in Borneo by Wallace and now preserved in The Hunterian, University of Glasgow (specimen numbers GLAHM 150611–613). (Photo by Dr Geoff Hancock, courtesy of The Hunterian).

PLATE. 7. The Sulawesi Ground-Dove (*Gallicolumba tristigmata*) inhabits the drier forests near Macassar and Manado, but Wallace found them to be quite rare. He singled them out to illustrate his paper on the pigeons of the Malay Archipelago (*Ibis*, vol. 1, 1865, plate 9, drawn by Joseph Wolf).

*11.*

PLATE. 9. (*left*) Wallace obtained a young female of the Northern Cuscus (*Phalanger orientalis*) on the Aru Islands. This animal with silvery hairs on the back was bought by the British Museum and listed by the Curator, John Edward Gray (*Proceedings of the Zoological Society of London* 1858, pl. 61, drawn by Joseph Wolf).

PLATE. 10. (*right*) Wallace's Scrubfowl or Moluccan Megapode (*Eulipoa wallacei*) from East Gilolo. When the specimen arrived in the British Museum, it was described by George Robert Gray, who mentioned 'the variability of its coloration, a peculiarity which imparts much interest to this new discovery of Mr. Wallace' (*Proceedings of the Zoological Society of London*, 1860, pl. 71).

PLATE. 11. Magnificent Bird-of-paradise (*Cicinnurus magnificus*) found near Dorey, New Guinea and in Mysool. The bird remained rare in European collections as it was seldom caught. This plate was drawn by John Gould and William Matthew Hart for Richard Bowdler Sharpe's *Monograph of the Paradiseidae* (1891–98), pl. 32.

PLATE. 12. Longicorn beetles attracted much of Wallace's attention. Many of them were described by Francis Polkinghorne Pascoe in London. This plate shows *Omotagus lacordairii* (from Dorey, 7 cm long), *Osphryon adustus* (from Dorey, 4 cm) and *Xaurus depsarius* (from Morty, 4 cm). (*Transactions of the Entomological Society of London*, vol. 3, 1864–69, pl. 22).

PLATE. 13. Birds of Paradise were greatly sought after by collectors for the colourful feathers of the male birds. In some cases the birds were assembled in great displays, like this one in the Bristol Museum of Natural History. (Photo by Dan Wye, www.dantwye.co.uk).

PLATE. 14.    Male Greater Bird of Paradise, by Travies, *Dictionnaire D'Histoire Naturelle*, 1849.

PLATE. 15.   The Austrian amateur entomologists Hans and Cajetan Felder obtained one of Wallace's specimens of Wallace's Golden Birdwing (*Ornithoptera croesus*). It was found on the island of Batchian and considered one of the most valuable butterflies. Wallace was the first to describe it, remarking that the 'beauty and brilliancy of this insect are indescribable.' (*Wiener Entomologische Monatsschrift*, vol. 3, 1859, pl. 7).

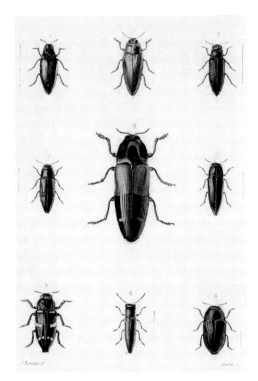

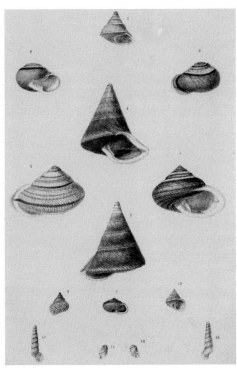

PLATE. 16.   A selection of jewel beetles (family Buprestidae) illustrating a paper on specimens collected by Wallace. These brilliant beetles were greatly prized by collectors despite their small size, and the ones shown were obtained by the Belgian entomologist Henri Deyrolle (*Annales de la Societe Entomologique de Belgique*, vol. 8, 1864, pl. 2).

PLATE. 17.   Wallace also collected shells as these too were part of the commercial collecting market. The two small specimens at bottom left and right (nos. 13–14) represent *Truncatella wallacei* in the collection of Saunders, collected in Waigiou. (*Transactions of the Zoological Society of London*, 1865, pl. 21).

PLATE. 18. (*above*)  Bantimurung Waterfall near Maros in Celebes, 1883. Watercolour by Josias Cornelis Rappard (1824–1889). (Tropenmuseum, Amsterdam)

PLATE. 19.  Wallace obtained a Siamang in Sumatra, which he took with him to Singapore. (William Jardine, *The Natural History of Monkeys*, 1833, pl. 4 *Hylobates syndactyla*).

PLATE. 20.  View of Ternate in the 1850s. (Steven A. Buddingh, *Neêrlands-Oost-Indië*, 1860, vol. 2, p. 77).

## 33. From Charles Darwin, 1 May 1857

**Down Bromley Kent**

*May 1.—1857*

My dear Sir

I am much obliged for your letter of Oct. 10th from Celebes received a few days ago: in a laborious undertaking sympathy is a valuable & real encouragement. By your letter & even still more by your paper in Annals,[187] a year or more ago, I can plainly see that we have thought much alike & to a certain extent have come to similar conclusions. In regard to the Paper in Annals, I agree to the truth of almost every word of your paper; & I daresay that you will agree with me that it is very rare to find oneself agreeing pretty closely with any theoretical paper; for it is lamentable how each man draws his own different conclusions from the very same fact.

This summer will make the 20th year (!) since I opened my first note-book, on the question how & in what way do species & varieties differ from each other.—I am now preparing my work for publication, but I find the subject so very large, that though I have written many chapters, I do not suppose I shall go to press for two years.—

I have never heard how long you intend staying in the Malay archipelago; I wish I might profit by the publication of your Travels there before my work appears, for no doubt you will reap a large harvest of facts.—I have acted already in accordance with your advice of keeping domestic varieties & those appearing in a state of nature, distinct; but I have sometimes doubted of the wisdom of this, & therefore I am glad to be backed by your opinion.—I must confess, however, I rather doubt the truth of the now very prevalent doctrine of all our domestic animals having descended from several wild stocks; though I do not doubt that it is so in some cases.—I think there is rather better evidence on the sterility of Hybrid animals than you seem to admit: & in regard to Plants the

---

[187]  Wallace, Sarawak Law, 1855.

collection of carefully recorded facts by Kölreuter & Gaertner, (& Herbert)[188] is *enormous.—*

I most entirely agree with you on the little effects of "climatal conditions", which one sees referred to ad nauseam in all Books; I suppose some very little effect must be attributed to such influences, but I fully believe that they are very slight.—It is really *impossible* to explain my views in the compass of a letter on the causes & means of variation in a state of nature; but I have slowly adopted a distinct & tangible idea.—Whether true or false others must judge; for the firmest conviction of the truth of a doctrine by its author, seems, alas, not to be slightest guarantee of truth.—

I have been rather disappointed at my results in the Poultry line; but if you sh^d· after receiving this stumble on any curious domestic breed, I sh^d be very glad to have it; but I can plainly see that this result will not be at all worth the trouble which I have taken. The case is different with the domestic Pigeons; from its study I have learned much.—The Rajah[189] has sent me some of his Pigeons & Fowls & Cats skins from interior of Borneo, & from Singapore.— Can you tell me positively that Black Jaguars or Leopards are believed generally or always to pair with Black? I do not think colour of offspring good evidence.—Is the case of parrots fed on fat of fish turning colour, mentioned in your Travels?[190] I remember case of Parrot with, (I think,) poison from some Toad put into hollow whence primaries had been removed.[191]

---

[188]   Three botanists: Joseph Gottlieb Kölreuter (1733–1806), Karl Friedrich von Gärtner (1772–1850), and William Herbert (1778–1847).

[189]   Sir James Brooke.

[190]   Darwin refers to Wallace, A.R. 1853. *Narrative of travels on the Amazon and Rio Negro*. London: Reeve & Co., pp. 321–2: 'the Mariánna [parrot]…was a most omnivorous feeder, eating rice, farinha, every kind of fruit, fish, meat, and vegetable, and drinking coffee too'. Wallace made no comment about change of colour.

[191]   Darwin, C. 1874. *Descent of man*, 2nd ed., London: John Murray, p. 60: 'It is also well to reflect on such facts, as the wonderful growth of galls on plants caused by the poison of an insect, and on the remarkable changes of colour in the plumage of parrots when fed on certain fishes, or inoculated with the poison of toads; for we can thus see that the fluids of the system, if altered for some special purpose, might induce other changes'.

One of the subjects on which I have been experimentising & which cost me much trouble, is the means of distribution of all organic beings found on oceanic islands; & any facts on this subject would be most gratefully received:

Land-Molluscs are a great perplexity to me.—This is a very dull letter, but I am a good deal out of health; & am writing this, not from my home, as dated, but from a water-cure establishment.[192]

With most sincere good wishes for your success in every way I remain My dear Sir

Yours sincerely

Ch. Darwin

# Macassar

## 34. To Henry Norton Shaw, August 1857

**Macassar**

*August 1857*

My dear Dr Shaw

You will receive from Mr. Stevens a paper of mine containing some account of the "Arru Islands,"[193] where I have spent six months, going & returning in a native trading prow. I went principally to shoot *Birds of Paradise*, & had capital sport. Of them I have sent an account to the "Annals of Nat. History."[194] If you are too full to read my paper perhaps Sir R. Murchison[195] would like it for the

---

[192]   Darwin was writing from Moor Park, Farnham, Surrey.

[193]   Wallace, A.R. 1858. On the Arru Islands. *Proceedings of the Royal Geographical Society of London* 2 (3): 163–70 [S41].

[194]   Wallace, A.R. 1857. On the Great Bird of Paradise [etc.]. *Annals and Magazine of Natural History* (ser. 2) 20 (120): 411–16 [S37].

[195]   Roderick Impey Murchison (1792–1871), a Scottish geologist, was president of the Geographical Society of London when Wallace wrote this letter, but was also an active member of the Geological Society of London.

"Geological" as it contains an account of some very curious & I think quite unique phenomenon of Physical Geography.

*This* is a horrid country, very bare & most difficult to get about. I leave as soon as I can "en route" for N. Guinea, about which I got much information at Arru. Having just got seven months letters & news I have plenty to do, besides extensive collections of birds & insects to sort & pack.

Trusting you are quite well, I remain,

My dear Dr. Shaw

Yours sincerely

Alfred R Wallace.

## 35. To Charles Darwin, [27 September 1857][196]

[Text excised] of May last, that my views on the order of succession of species were in accordance with your own, for I had begun to be a little disappointed that my paper had neither excited discussion nor even elicited opposition. The mere statement & illustration of the theory in that paper is of course but preliminary to an attempt at a detailed proof of it, the plan of which I have arranged, & in part written, but which of course requires much research in English libraries & collections, a labour which I look [text excised].

With regard to the black Jaguars always breeding *inter se*, it is of course a point not capable of proof, but the black & the spotted animals are generally confined to separate localities, & among the hundreds & thousands of the skins which are articles of commerce I have never heard of a parti-coloured one having occurred. I *think* there is a difference of form the black being the more slender & graceful animal. [text excised]

---

[196]  Darwin annotated the fragment: '(Alfred R Wallace. Letter Sept. 1857.)' The full date is included in Darwin's letter to Wallace of 22 December 1857.

# 36. From James Brooke, 31 October 1857

Singapore

*31ˢᵗ Octʳ 1857*

My dear Wallace

Before I quit Singapore for England I must tell you that your papers on the Orangs[197] turned up, after my letter had been dispatched. I agree in your conclusion of the ridge on the cranium being attributable to individual peculiarities, but on the whole before coming to any positive conclusion on the number of species it will be advisable to wait for more facts and a larger field of investigation. Your collection was made in a limited space of country and should not therefore be held conclusive as ~~appli~~ settling the general question—Here for instance is the measurement of a Mias killed by my nephew C. Johnson,[198] with his remarks—

"September 3ʳᵈ 1857.

Killed female Orang Utan. Height, from head to heel 4ᶠᵗ 6ⁱⁿ.

Stretch from finger to finger across body, 6[ᶠᵗ]1[ⁱⁿ]

Breadth of face (including callosities) 11[ⁱⁿ]

Callosities large—middle aged animal—hair reddish brown, thick and shaggy, with beard, and hair on the upper lip."[199]

He added that the callosities were hard, ie as hard as the gristle of a man's nose. — The head I will direct to be sent home when cleaned. This points from size alone to a distinct species—however I have not time to discuss.—

I wish you very well and happy,—I am neither ~~very~~ the one or the other at present and suffering the consequences of follies long passed—Change I hope will do me good and if England suits me not I shall take refuge on the continent. —

---

[197]  Wallace, A.R. 1856. On the Orang-utan or Mias of Borneo. *Annals and Magazine of Natural History* (ser. 2) 17 (102): 471–6 [S24].

[198]  Charles Anthoni Johnson Brooke (1829–1917), who succeeded James Brooke as Rajah of Sarawak.

[199]  Wallace quoted these details in MA1: 98.

Write me whenever you feel inclined you know it affords me pleasure to hear from you and of your success.

Believe me Yours very sincerely

J Brooke.

A Wallace Esq.

## 37. To Samuel Stevens, 20 December 1857

**Amboyna**

*December 20, 1857.—*

My collecting this year has been so peculiar and so different from anything I have yet done in the tropics that I must give you some little account of it; my locality was at the foot of the mountains about thirty miles north of Macassar,[200] the whole country between this range and the sea is a dead level of paddy fields, flooded for half the year, and of course absolutely barren of insects; the mountains are of limestone or basalt, the former rising from the plain in immense perpendicular walls quite inaccessible, except where a few streams break through them; the basalt hills are more rounded, and at the foot of one of them is a forest of palms and jack fruit. I had a small bamboo house built; when I arrived in August there had not been rain for two months and it was fearfully hot and parched; dead leaves strewed the ground, and a beetle of any kind was sought for in vain.

After some time I found a rocky river-bed issuing from a cleft in the mountains, and though dry it still contained a few pools and damp hollows; these were the resort of numerous butterflies,—Papilio Euryphilus, the new species near Sarpedon, P. Rhesus, P. Peranthus and the rare P. Encelades, Bois., the beautiful Pieris Zaranda was rather abundant, and several interesting

---

[200] Wallace stayed near the village of Maros.

Nymphalidæ. Here, therefore, I made daily excursions and procured good series
of many of these insects; the paths in the forest adjoining this stream were
pretty abundant in Ornithoptera; of two species, O. Remus and the very rare
O. Haliphron, *Bois.*, both sexes of which I took, and twice *in copulâ*; the female
something resembles O. Amphimedon, which is the female of O. Helena.
About the mud holes Hymenoptera were abundant and on the fallen palm
stems; in dry gulleys, &c. were many very curious Diptera; Coleoptera, however, were not to be found: I searched dead trees, and bark and leaves, with no
other reward than a very few species of minute Curculios and obscure
Chrysomelidæ.

After a few weeks of this work the mud holes got baked hard, the pools of
water disappeared one after another, and with them the butterflies and other
insects, and for some days I got almost nothing. I now set to turning over the
stones and dead leaves in the sandy river-bed, and soon found that there were
some minute Coleoptera under them, namely, Anthici and very small Carabidæ; to catch them I made my boy bring a basin of water and a spoon, and by
shoveling in the sand I could pick off the insects which floated on the surface:
in this way I got many Carabidæ, the largest not more than 1½ line; two or
three species of Anthicus and some Steni and other Brachelytra. I now turned
my attention to buffalo-dung, which, though very barren compared with genuine
British cow-dung, would I found yield something to a persevering search,—
I obtained Histers, Onthophagi, and a considerable number of minute Staphylinidæ. A few days, however, soon exhausted this collecting-ground, for,
except in the river-bed, the dung was absolutely uninhabited, when chance
showed me a new and very rich beetle station.

My lad brought me one day a fine large Nitidula which he had found in an
over-ripe jack fruit (*Artocarpus* sp.); this set me to searching these fruits, of
which there were a number about in various stages of decay, and I soon found
that I had made a discovery,—Staphylinidæ, large and small, Nitidulæ,
Histers, Onthophagi, actually swarmed on them: every morning, for some
weeks, I searched these rotten fruits, and always with more or less success;

I placed ripe ones on the fruit here and there, which I visited once a day, and from some of them got even Carabidæ; in all I found not much short of one hundred species of Coleoptera on the fruit, including most that I had before found in dung, so that it seems probable that, in tropical countries, the large fleshy fruits in a state of decay and putrescence are the true stations of many of the Carpophagous and Necrophagous Coleoptera, a fact of some importance, as explaining the presence of Onthophagi, &c. in places where there are no ruminating animals: at length the rains began to fall almost every evening, and the fruits, soaked with water, ceased to be productive, but I was compensated by discovering that the margins of the streams, which when dry were so rich in Lepidoptera, were now an excellent collecting-ground for small Coleoptera; under the moist dead leaves that lay on the rocks I found numbers of small and very interesting Carabidæ, with hosts of Anthici, and a good many Pselaphidæ and Hydrophili: with the rains the butterflies almost disappeared, while the Cicindelidæ came out in great abundance, four species being different from those I took last year; small Melolonthidæ also now became abundant on the foliage, and I took two or three species new to me, with several pretty Chrysomelas and Curculios. After a fortnight's close work at minute Coleoptera, the weather became so wet and cloudy, as to admonish my return to Macassar to pack my collections before the commencement of the continuous heavy rains.

To persons impressed with the idea of the prevalence of large insects in the tropics, my Macassar collections will appear most extraordinary; the average size is certainly less than that of our British species, and the colours not at all more brilliant. Of the Carabidæ (more than one hundred species), the greater part are under 4 lines and a very large number under 2 lines, whilst several under 1 line[201] are perhaps the smallest of the family: the Brachelytra (eighty or ninety species) are, with the exception of about a dozen, very minute and obscure: the Rhynchophora are all small, and there are about one hundred species of minute

---

[201]    A foot (30.48 cm) is divided into 12 inches (2.54 cm) of 12 lines (0.21 cm).

Necrophaga, Xylophaga, &c., and about eighteen species of the elegant little Anthici, whilst the Longicornes, Buprestidæ and Cetoniæ, usually so abundant, are very scarce: if we were to take away some dozen purely tropical forms, the collection would have all the appearance of one from an extratropical and even northern locality, owing to the large proportion of Carabidæ, Staphylinidæ and Necrophaga, the small average size of the species and the obscurity of their colours.

Amboyna, where I am staying a month only, on my way to Ternate, offers a striking contrast to the country I have just quitted: it is eminently tropical; the number of large and handsome species in all orders of insects is perhaps greater than in any other place I have visited, and the forms far more closely resemble those of Aru than of Borneo or Macassar; a number of the common species of the surrounding island are represented at Amboyna by others very closely allied or by varieties, but in almost every instance they are of larger size and more brilliant colours,—Papilio Severus and Ulysses are larger here than at Aru, whilst Deiphobus is larger than the closely allied Memnon of the Sanda Island or Ascalaphus of Macassar.

In the Hymenoptera, the species of Vespidæ and Pompilidæ are gayer than the allied species I have found in other countries; a Laphria and an Anthrax are larger than any Diptera I have yet found of the same genera; while the Coleoptera include the gigantic Eucheirus longimanus and a number of large and handsome Longicornes, Buprestidæ and Anthribidæ: it may be easily imagined, therefore, that Amboyna is a tempting place, well worth a thorough exploration, and I shall probably return to it unless I shall be able to visit Ceram, which I expect will contain almost all the Amboyna species, and probably many more, as is known to be the case with the birds. Though everybody says this is the dry and hot season, yet the weather has been terribly wet and windy, and during the twelve days I have now resided in a little hut in the jungle I have not had a single hot sunny day; here, as everywhere in the East, there is no forest left for many miles round the town, and there was the usual difficulty in finding a locality and a home, and in conveying my baggage.

In the town I reside with Dr. Mohnike,[202] the chief physician of the Moluccas, a German, an entomologist, and a very learned and hospitable man; he has lived in Japan, made a voyage to Jeddo, ascended volcanoes, and made collections: my pleasure may be imagined in looking over his superb collection of Japanese Coleoptera, large and handsome Longicornes and Lucani, tropical Buprestidæ and northern Carabi: he has also an extensive collection of Coleoptera made during many years' residence in Sumatra, Java, Borneo, and the Moluccas—a collection that makes me despair; such series of huge Prioni, Lamiæ and Lucani, Dynastidæ and Eucheirus! It is such collections that give, and have always given, such an erroneous idea of Tropical Entomology: these collections are made entirely by natives. Dr. Mohnike has resided here in Amboyna, for example, *two* years, and every native in the island knows that large and handsome beetles will be purchased by him; he has, therefore, hundreds of eyes spread over hundreds of square miles, and thus species which in ten years might never once occur to a single collector, are inevitably obtained by him in greater or less abundance, whilst the smaller, more active, and much more common species are never brought at all.

The Eucheirus is evidently rare, yet Dr. Mohnike has a fine series, obtained at intervals from different localities; he also sends bottles and casks of arrack to the Dutch officers resident in different islands, and though he sometimes has them returned crammed full of a single species of common Calandra or Passalus, yet he occasionally gets some magnificent insects. I believe myself that, as a general rule, beetles are rare exactly in proportion to their size, rare both in species and in individuals; in four years' almost daily search in the Eastern forests I have never found a large Prionus myself, and I have collected nearly four thousand species of Coleoptera: such collections as those of Mr. [Henry Walter]

---

[202]   Otto Gottlieb Johan Mohnike (1814–1887), German physician, entomologist, and naturalist and chief medical officer of the Dutch Moluccas. Mohnike later published an important book on the natural history of the Dutch East Indies: Mohnike, O.G. 1883. *Blicke auf das Pflanzen- und Thierreich in den niederländischen Malaienländern*. Münster: Aschendorff'sche Buchhandlung.

Bates and myself, made in such distant countries (both generally considered among the richest in large species), are what show the true nature of tropical insects, and I believe that a careful examination of these will lead to the conclusion that there is no superiority whatever in the average size of tropical Coleoptera over those of temperate climates, and that in many groups the latter have the decided advantage.

A. R. Wallace.

## 38. From Charles Darwin, 22 December 1857

**Down Bromley Kent**

*Dec. 22/57*

My dear Sir

I thank you for your letter of Sept. 27[th.] I am extremely glad to hear that you are attending to distribution in accordance with theoretical ideas. I am a firm believer, that without speculation there is no good & original observation. Few travelers have attended to such points as you are now at work on; & indeed the whole subject of distribution of animals is dreadfully behind that of Plants. You say that you have been somewhat surprised at no notice having been taken of your paper in the Annals:[203] I cannot say that I am; for so very few naturalists care for anything beyond the mere description of species. But you must not suppose that your paper has not been attended to: two very good men, Sir C. Lyell[204] & Mr E. [Edward] Blyth at Calcutta specially called my attention to it. Though agreeing with you on your conclusions in that paper, I believe I go much further than you; but it is too long a subject to enter on my speculative

---

203   Wallace, Sarawak Law, 1855.
204   Charles Lyell (1797–1875), Scottish geologist and author of the most important work for Wallace's evolutionary theorizing: *The principles of geology*, 4th ed., 1835.

notions.—I have not yet seen your paper on distribution of animals in the Arru Isl^{ds}:[205]—I shall read it with the <u>utmost</u> interest; for I think that the most interesting quarter of the whole globe in respect to distribution; & I have long been very imperfectly trying to collect data for the Malay archipelago.—

I shall be quite prepared to subscribe to your doctrine of subsidence: indeed from the quite independent evidence of the Coral Reefs I coloured my original map in my Coral volume[206] of the Arru Isl^{d.} as one of subsidence, but got frightened & left it uncoloured.—But I can see that you are inclined to go <u>much</u> further than I am in regard to the former connections of oceanic islands with continents: Ever since poor E. Forbes[207] propounded this doctrine, it has been eagerly followed; & Hooker[208] elaborately discusses the former connections of all the Antarctic isl^{ds} & New Zealand & S. America. About a year ago I discussed this subject much with Lyell & Hooker (for I shall have to treat of it) & wrote out my arguments in opposition; but you will be glad to hear that neither Lyell or Hooker thought much of my arguments: nevertheless for once in my life I dare withstand the almost preternatural sagacity of Lyell.—

You ask about Land=shells on islands far distant from continents: Madeira has a few identical with those of Europe, & here the evidence is really good as some of them are sub-fossil. In the Pacific isl^{ds} there are cases, of identity, which I cannot at present persuade myself to account for by introduction through

[205] Wallace, A.R. 1858. On the Arru Islands. *Proceedings of the Royal Geographical Society of London*, 2 (3): 163–70 [S41].

[206] Darwin, C.R. 1842. *The structure and distribution of coral reefs*. London: Smith Elder and Co., plate 3. The Aru islands remained uncoloured in the second edition of 1874.

[207] Edward Forbes (1815–1854), zoologist, botanist, and palaeontologist. Browne, Janet. 1983. *The secular ark: studies in the history of biogeography*. New Haven and London: Yale University Press, pp. 144–54.

[208] Joseph Dalton Hooker (1817–1911): botanist; assistant director, Royal Botanic Gardens, Kew, 1855–1865; director, 1865–1885. He wrote about the connections of the Antarctic islands in the introductory essay of *Flora Novae-Zelandiae*. London: Lovell Reeve, 1853.

man's agency; although D$^r$ Aug. Gould[209] has conclusively shown that many land-shells have there been distributed over the Pacific by man's agency. These cases of introduction are most plaguing. Have you not found it so, in the Malay archipelago? it has seemed to me in the lists of mammals of Timor & other islands, that *several* in all probability have been naturalised.

Since writing before, I have experimentised a little on some land-mollusca & have found sea-water not quite so deadly as I anticipated. You ask whether I shall discuss "man";—I think I shall avoid whole subject, as so surrounded with prejudices, though I fully admit that it is the highest & most interesting problem for the naturalist.—My work, on which I have now been at work more or less for 20 years, will *not* fix or settle anything; but I hope it will aid by giving a large collection of facts with one definite end: I get on very slowly, partly from ill-health, partly from being a very slow worker.—I have got about half written; but I do not suppose I shall publish under a couple of years. I have now been three whole months on one chapter on Hybridism!

I am astonished to see that you expect to remain out 3 or 4 years more: what a wonderful deal you will have seen; & what interesting areas, the grand Malay Archipelago & the richest parts of S. America!—I infinitely admire & honour your zeal & courage in the good cause of Natural Science; & you have my very sincere & cordial good wishes for success of all kinds; & may all your theories succeed, except that on oceanic islands; on which subject I will do battle to the death.

Pray believe me,

My dear Sir

Yours very sincerely

C. Darwin

---

[209]   Augustus Addison Gould (1805–1866) was an American conchologist. Darwin refers to Gould, A.A. 1852–1856. *Mollusca and shells. Vol. 12 and atlas of United States Exploring Expedition during the years 1838–42.* Philadelphia.

# Amboyna

## 39. To Philip Lutley Sclater, [January 1858]

[Amboyna][210]

P.S. In my next voyage to New Guinea I think it probable I may get some live *Paradiseus (P. papuana)* but I must have a definite arrangement or will not trouble myself with them. I hear from Capt. of steamer there is one now in Batavia for which 1000 Rupees (£85) is asked. This is too much but it shows their value here. Now I myself will not come home on any chance, & if sent a person must come to take charge of them. If therefore the Crystal Palace Co. wants them, you must get & send me out an order for a free passage from Singapore to England, first class to any person in charge of Birds of Paradise for me.

Fig. 16. Philip Lutley Sclater

Next they must either be put up to auction on arrival & the Palace get them at their market price, or they must agree to pay as follows: if only one comes live £100, the 2nd £50, 3rd & others up to 10, £25 each. If they won't ~~offer~~ give this price I will not trouble myself even if I can get them for nothing. I must have their answer immediately & it must be understood that they take their chance of how many are females as in the young birds I cannot tell the difference.

---

[210] This is an extract of a previous communication, enclosed with letter to P.L. Sclater of 4 April 1862. The place Amboyna and the date are based on the mention of Dr Doleschall, whom Wallace met in Amboyna.

This is my ultimatum. I would rather sell them in Ternate for £5 each than trouble myself about them for less. Mind & take care of all the light sago-pith boxes in which I send my insects now, as they will be invaluable for me in arranging my coll^n of insects & birds, & make excellent show boxes for the large beetles & small birds. All Mr. Doleschall's collections here go to Vienna, & will lower the market a little, but few I think will be in good condition. In a few years E. longimanus will be very common as the Doctors at Amboina have got perhaps hundreds of them. Send me as much as you can information as to which of my large & handsome sp. are quite new.

I must now conclude again,

Yours very faithfully,

Alfred R. Wallace

## 40. To Henry Walter Bates, 4 and 25 January 1858

**Amboyna**

*Jan 4th. 1858*

My dear Bates,

My delay of six months in answering your very interesting & most acceptable letter dated "Tunantins, 19 Nov. /56" has not I assure you arisen either from laziness or indifference, but really from pressure of business & an unsettled state of mind. I received your letter at Macassar on my return in July last from a seven months' voyage & residence in the Arru Islands close to New Guinea. I found letters from Australia, from California, from you, from *Spruce*, from Darwin, from home, & a lot of interesting *Stevensian* dispatches.

I had 6 months' collections (mostly in bad condition owing to dreadful dampness & sea air) to examine & pack; about 7 thousand insects having to be gone over individually & many of them thoroughly cleaned; besides an extensive collection of birds. I was thus occupied incessantly for a month, & then immediately left for a new locality in the interior, where I staid 3 months during which time I had most of my correspondence to answer & was besides making some collections so curious & interesting that I did not feel inclined to answer your letter till I could tell you something about them. At the end of October I returned to Macassar, packed up my collections & left by steamer for Ternate, viâ this place, where I have staid a month, had some good collecting & it is now, on the day of my departure, having all my boxes packed & nothing to do, that I commence a letter to you.

Your letter has been a source of much pleasure & interest to me. I have read it & reread it at least 20 times. In particular, your list of species is most interesting to me, only I wish you had made up a complete list supplying the Pará species &c. by conjecture. In your *Coleoptera* the only thing that really astonishes me & for which I was not at all prepared is your vast number of *Carabidae* [ground beetles]. It is the group in which you *most* decidedly surpass me (not reckoning the *Erotyli* which are almost peculiar to America). In *Cicindelae* [tiger beetles] we are about equal, that is comparing my 3½ years coll$^g$ with the 5 years of which you have given me your statistics. In *Cleri* [checkered beetles] you also decidedly surpass me as I do not think I have much above 50 species. Of Longicorns I have *now* about 550 species & they will average, I think, a little larger than yours. Your *Prionidae* & *Cerambycidae* will be I think more varied & beautiful than mine; but my *Lamiae* are much the most numerous & contain some superb species. My *Lamiae* in fact form near 4/5 of all my Longicorns & are nearly 3 times my number of Cerambycidae, while with you the two groups are not very unequal.

In *Rhyncophora* [weevils] again I have now near a thousand species, swarms of minute & obscure things of course, but also a number of very beautiful Anthribidae & Brenthidae. Our Lamellicornes [beetles] are nearly equal, but I surpass you in *Cetonias* & Lucanidae while you have the great superiority in your *Copridae*. The only handsome group in which I think I shall be decidedly the best off is the Buprestidae of which I have perhaps 150 species of which 60 are above ½ in. & many very brilliant.

In the *Elateridae* [click beetles] we are about equal, but in the *Cyclica* you considerably surpass me in your number of species. At Macassar this year I made an extraordinary collection of minute things. *Anthici* I think about 18 species, many *Pselaphidae* & hosts of minute & obscure *Philhydrida* & *Necrophaga*. For the first time too I really collected *Staphylinidae* [rove beetles], getting many species under dung in sand but *principally* in <u>rotten fruits</u>. As near as I could make out I got 90 species mostly very minute, so that I think it is pretty evident there are plenty of them in the tropics but hitherto no one has had time or inclination to search for them but *you, myself* & *J.C. Bowring* who in Hongkong has taken also *92 species*! The *rotten fruit* of a large fleshy *Artocarpus* (Jackfruit) was wonderfully productive, for besides the Staphylinidae I took fine *Nitidulae* & *Sphaeridii* in it, about a dozen species of <u>Onthophagus</u>, & two <u>Carabidae</u>!! These last howeve, I took most abundantly after the commencement of the rains by beating dead leaves & under decaying leaves on the [word crossed out] rocky margins of mountain streams. I got thus hosts of curious *Brachinidae* & *Harpalidae*, mostly very small, some indeed are I think the smallest *Carabidae* known as I have several species under a line. All this time however you must remember I was getting nothing that can be called *fine* in Coleoptera,—no Longicorns, the minutest of Curculionidae, no Buprestidae or Lucani.

In Lepidoptera [butterflies] however I enjoyed the luxury of capturing four species of *Ornithoptera*, the largest number yet known to exist in one locality. One which I took most abundantly was I believe unique in Europe (*O. haliphrous* Bois.). I also took sparingly the grandest known swallow tail (*Papilio androcles*, Bois.) near twice as big as *protesilaus*! I was also making at the same time very fine collections of *Hymenoptera* [wasps, bees] & *Diptera* [flies] (*270* & *202* species). Of the number of *Dip:* you may form some idea by the fact that after having taken 140 species, I took 30 new ones in one day! I did of course little else as I had found a fine station for them near my house in the forest;—soon after I took to the minute Col^a & of course neglected the poor dipt^a, so you may imagine how numerous the species are here. I think I have now collected 600 species of Eastern diptera & hope to reach 1000 before my return which I expect is about as many as all the Exotic dip. previously known.

Here in Amboyna I have had 20 days collecting & have taken 290 species of coleoptera, 58 are Longicorns containing some fine things, & one perhaps the handsomest species I possess, a *Monohammus*, about 1½ in. long, blue black with broad bands of a dense pubescent orange buff. I also procured a few of the grand *Euchirus longimanus*, and a series of most beautiful *Buprestidae*.

The true *Priamus* I did not see, but the gorgeous *Ulysses*, the prince of Papilios, is not uncommon, & I got several fine specimens. I enjoyed the society of two Entomologists, the government doctors, a German *Dr. Mohnike* with whom I lived, a schoolfellow of Burmeister & Erichson.[211] He has been to Japan & made there a nice collection of perhaps 300-400 sp. of Coleoptera. He gave me about 50 species of his duplicates. *D*ʳ *Doleschall*[212] is a Hungarian who studied a year in the Vienna Museum (the *Diptera* & *Arachnida*) which he knows well. He also collects the Lepidop. & Colᵃ of Amboyna, and liberally gave me a fine suite for my private collⁿ. He is a delightful young man, but poor fellow's dying of consumption. He can hardly I fear live a year, yet is enthusiastic in Entomology. He says Hungary is very rich in Coleoptera & contains about a *hundred species* of *true Carabus*! Talk of the tropics after that! We conversed always in French, of which I have had to make so much use that I am getting tolerably fluent though fearfully ungrammatical. But we were about equal in that respect & so blundered along gloriously.

To persons who have not thought much on the subject I fear my Paper "On the Succession of Species"[213] will not appear so clear as it does to you. That paper is of course merely the announcement of the theory, not its development. I have

---

[211] In Stralsund, Otto Mohnike grew up with Karl Hermann Konrad Burmeister (1807–1892) and Wilhelm Ferdinand Erichson (1809–1848), both known for entomological research.

[212] Carl Ludwig Doleschall (1827–1859), Hungarian physician and entomologist. See Stagl, V. 1999. Carl Ludwig Doleschall: Arzt, Forscher und Sammler. *Quadrifina* 2: 195–203.

[213] Wallace, Sarawak Law, 1855.

prepared the plan & written portions of an extensive work embracing the subject in all its bearings & endeavouring to prove what in the paper I have only indicated. It was the promulgation of *"Forbes's theory"*[214] which led me to write & publish, for I was annoyed to see such an ideal absurdity put forth when such a simple hypothesis will explain *all the facts*.

I have been much gratified by a letter from Darwin,[215] in which he says that he agrees with *"almost every word"* of my paper. He is now preparing for publication his great work on *Species & Varieties*, for which he has been collecting information 20 years. He may save me the trouble of writing the 2nd part of my hypothesis, by proving that there is no difference in nature between the origin of species & varieties, or he may give me trouble by arriving at another conclusion, but at all events his facts will be given for me to work upon. Your collections & my own will furnish most valuable material to illustrate & prove the universal applicability of the hypothesis.

The connection between the succession of affinities & the geographical distribution of a group, worked out species by species, has never yet been shown as we shall be able to show it. In this Archipelago there are two distinct faunas rigidly circumscribed, which differ as much as those of S. Am. [South America] & Africa, & more than those of Europe & N. Am. yet there is nothing on the map or on the face of the islands to mark their limits. The Boundary line often passes between islands closer than others in [word crossed out] the same group. I believe the W. part to be a separated portion of continental Asia, the Eastern, the fragmentary prolongations of a former Pacific continent. In Mammalia & birds the distinction is marked by *genera, families*, & even *orders* confined to one

---

[214]  Edward Forbes. There are at least two articles by Forbes that Wallace may have read, but he did not record the precise reference and textual clues are not conclusive, i.e. Forbes, E. 1854. Anniversary address of the President (17 February 1854). *Quarterly Journal of the Geological Society of London* 10: xxii–lxxxi; and Forbes, E. 1854. On the manifestation of polarity in the distribution of organized beings in time. *Notices of the Proceedings of the Royal Institution of Great Britain* 1: 428–33.

[215]  Darwin's letter of 1 May 1857.

region,—in Insects by a number of genera & little groups of peculiar species, the *families* of insects having generally a universal distribution.

*Ternate, Jan. 25th*

I have not done much here yet, having been much occupied in getting a house repaired & put in order. This island is a volcano with a sloping spur on which the town is situated. About 10 miles to the E. is the coast of the large island of *Gilolo* perhaps the most perfect Entomological "terra incognita" now to be found. I am not aware that a single insect has ever been collected there, & cannot find it given as the locality of any insects in my catalogues or descriptions. In about a week I go for a month collecting there, & then return to prepare for a voyage to N. Guinea. I think I shall stay in this place 2 or 3 years, as it is the centre of a most interesting & almost unknown region. Every house here was destroyed in 1840 by an earthquake during an eruption of the volcano. The Dutch steamer comes here every month & brings letters from England in about 10 weeks which makes the place convenient & there are also plenty of small schooners & native Prows by which the surrounding islands can be visited.

What great political events have passed since we left England together! And the most eventful for England & perhaps the most glorious, is the present mutiny in India which has proved British courage & pluck as much as did the famed battles of Balaclava & Inkermann. I believe that both India & England will gain in the end by the fearful ordeal. When do you mean returning for good? If you go to the Andes I think you will be disappointed, at least in the *number of species*, especially of Coleoptera. My experience here is that the low grounds are *much* the most productive, though the mountains generally produce a few striking & brilliant species. I must now conclude wishing you a safe return to England.

Yours sincerely
Alfred R Wallace.

*W.H. Bates Esq.*

*Postscript.*

I have just taken my first true *Pachyrhyncus*, a genus of remarkably restricted range.

There are many other topics on which I have not space to touch. I trust the day may come when both returned home, we may visit each other, compare our collections, & discuss those questions we both find of so much interest. There are many hitherto untouched branches of enquiry in entomology which our collections & statistics will enable us to develop. I see occupations for a life of delightful study. May we both live to realise it!

ARW

[enclosure]

Insects of the Eastern Archipelago, collected by AR. Wallace from May 1, 1854 to January 1st 1858.

| Coleoptera | Sing.ᵉ & Malac. 6 months | Borneo 15 months | Macassar 6 months | Aru Is. 6 months | Amboyna. 1 month | Probable totals. |
|---|---|---|---|---|---|---|
| Geodephaga . | 150 | 70 | 125 | 20 | 16 | 220 |
| Hydrophili.... | 4 | 7 | 28 | 3 | 0 | 35 |
| Staphylini.... | 15 | 26 | 90 | 6 | 1 | 120 |
| Necrophaga | 25 | 50 | 120 | 20 | 6 | 180 |
| Histor &c &c .. | | | | | | |
| Lamellicornes ... | 55 | 72 | 70 | 18 | 9 | 175 |
| Pectinicornes ... | 16 | 37 | 2 | 9 | 4 | 55 |
| Buprestidae ... | 24 | 100 | 30 | 23 | 25 | 170 |
| Elateridae..... | 45 | 125 | 38 | 30 | 14 | 200 |
| Longicornes ... | 160 | 290 | 38 | 114 | 58 | 570 |
| Rhyncophora .. | 245 | 570 | 135 | 145 | 85 | 950 |
| Malacoderma ... | 60 | 100 | 21 | 45 | 12 | 190 |
| Cleridae.... | 30 | 41 | 13 | 16 | 10 | 70 |
| Heteromera ... | 75 | 140 | 70 | 42 | 20 | 240 |
| Cyclica..... | 160 | 250 | 92 | 68 | 22 | 450 |
| Trimera.... | 15 | 50 | 16 | 13 | 8 | 75 |
| | 979 | 1928 | 898 | 572 | 290 | 3700 |

The above totals are I think under, rather than over the mark.

| Lepidoptera . . | Total no. of *species* (*estimate*) |
|---|---|
| Papilionidae . . . . . | 58 |
| Pap. 50, Ornith.7, Lept.1 | |
| Pieridae . . . . . | 70 |
| Danaidae . . . . . | 40 |
| Nymphalidae &c . . . . . | 160 |
| Satyridae . . . . . | 32 |
| Lycaenidae &c . . . . . | 180 |
| *Hesperidae* . . . . . | *80* |
| Total Diurni . . . . . | 620 |
| | |
| Other Lepidoptera . . . . . | abᵗ 2000 |
| Hymenoptera . . . . . | 750 |
| Diptera . . . . . | 660 |
| Hemiptera . . . . . | 340 |
| Homoptera . . . . . | 160 |
| Orthoptera . . . . | 160 |
| Neuroptera . . . . . | 110 |
| Dermap. Forfic. [Forficulae] &c ... | 40 |
| *Coleoptera* . . . . . | *3700* |
| Total species . . . . . | *8540 of Insects* ... |

The numbers here given are certainly not far from the truth. It appears that in all orders except *Coleop.* & *Lep. diurnes* I surpass you considerably in the number of species. Whether it is owing to your attention being so much absorbed by the hosts of your butterflies as to prevent your attending to the less interesting groups, or whether it shows a real deficiency in S. America to balance its undoubted superiority in *Diurni*, is a question of much interest. You might partly solve it by resolutely devoting a month to the Hymenop. Diptera &c. almost exclusively, which compared with my average months collecting in these groups would furnish data for determining the point.

ARW.

# TERNATE AND NEW GUINEA
## 8 January–8 October 1858

Wallace arrived on Ternate on 8 January 1858. He brought letters of introduction to a wealthy Dutch merchant named Maarten Dirk van Duivenbode (1805–1878) who helped Wallace rent a house from a Chinese man on the base of the volcano overlooking the small town. Within about a month Wallace was to conceive of natural selection as an explanation of evolutionary change and adaptation. However, as we have no contemporary records the course of events has remained rather mysterious.

On 25 January 1858 Wallace continued his letter to Henry Bates. Wallace mentioned the larger nearby island of Gilolo to the East, 'the most perfect Entomological "terra incognita" now to be found'. Wallace planned to go there in 'about a week...for a month collecting'. This remark was used by historian H. Lewis McKinney in 1972 to argue that Wallace must have left for Gilolo around 30 January. According to Wallace's *Journal,* he returned to Ternate on 1 March. Hence McKinney concluded that Wallace was on Gilolo during the entire month of February, and consequently that Wallace conceived of natural selection and wrote his famous essay on Gilolo, not on Ternate.[216] Most writers have since adopted this view. To explain why his essay was signed 'Ternate', it has been assumed that Wallace used it merely as his postal base or address. However, his correspondence does not support this otherwise quite reasonable proposal. Wallace seems always to have given his actual location, whether or not it possessed a post office. Although widely accepted, McKinney's conclusion was rather hasty. In fact, Wallace was still on Ternate during the first half of

---

[216] McKinney, H.L. 1972. *Wallace and natural selection.* New Haven: Yale University Press.

February—thus he conceived of natural selection in his house on Ternate as he always said. Here is how to disentangle the mystery.

It has long been assumed that Wallace carried out his plan mentioned to Bates to leave in about a week and to stay on Gilolo for 'a month'. However, three sources reveal that Wallace did not actually carry out this plan. Instead, he stayed on Gilolo only for two weeks in February 1858, not for a month. When adjusting dates of his journey on a small document preserved in Oxford, he noted that he stayed in Gilolo for two weeks.[217] In the 2 March 1858 letter to Frederick Bates, Wallace states that he received Bates' letter 'a month ago' — which would mean it arrived on the c. 8 February mail steamer at Ternate. There was no mail service on Gilolo. In addition, if Wallace was ready to depart around 1 February it seems likely that he would have waited for the monthly mail steamer, which was usually due between the 6th and 8th, because he needed to collect fresh supplies that were already a month overdue.[218] It is curious that no Wallace manuscripts or other documents bear a date in February 1858. His manuscript essay outlining natural selection is lost, but it was printed for the Linnean Society of London. The printed version was dated 'Ternate, February, 1858'. In all of his published recollections, Wallace described his discovery of natural selection taking place in his house on Ternate, when he was suffering from a recurrent bout of fever.[219]

---

[217] Oxford University Museum of Natural History Library, mentioned in Smith, A.Z. 1986. *A history of the Hope Entomological Collections in the University Museum Oxford with lists of archives and collections* (Oxford University Museum Publication 2). Also, Wallace, A.R. 1869. Notes on the localities given in Longicornia Malayana, with an estimate of the comparative value of the collections made at each of them. *Transactions of the Entomological Society of London* (ser. 3) 3 (part VII): 691–6. Page 695 mentions that he 'did not spend more than a month there' during his voyage. His second visit there was for two weeks.

[218] For a detailed discussion of this issue of February 1858, see Wyhe, J. van. 2013. *Dispelling the darkness* pp. 202–4.

[219] Wallace's recollections can be found in the following items: (1) Letter to A.B. Meyer, 22 November 1869, published in *Nature*, 52 (1895): 415; (2) Letter to Alfred Newton, 3 December 1887, in Darwin, F. ed. 1892. *Charles Darwin: his*

The fact that Wallace really was on Ternate in early February means that recent theories proposing that the human races on Gilolo inspired natural selection cannot be correct. Wallace thought of natural selection before visiting Gilolo. In fact, the best contemporary evidence we have as to what inspired Wallace's breakthrough is found in his correspondence.

Writing to Frederick Bates, the brother of Henry, on 2 March 1858, Wallace noted the striking match between the colours of certain tiger beetles with the sand or mud on which they were found. One of these species, collected at Macassar six months before, 'was found in the soft shiny mud of salt creeks, with which its colour so exactly agrees that it was perfectly invisible except for its shadow! Such facts as these puzzled me for a long time, but I have lately worked out a theory which accounts for them naturally'. This is obviously a reference to the theory in the Ternate essay, which twice mentions this colouration phenomenon.

In his previous evolutionary theorizing, Wallace never mentioned adaptation to local circumstances but focused instead on the order in which living types had appeared during the history of life, genealogical common descent, and preformed features that led to modifications in later descendants. Thinking of the tiger beetles, Wallace seems to have combined several threads of his earlier thinking. Textual clues in the Ternate essay make it clear that Wallace consulted his earlier theoretical notes in his *Notebook 4* while writing out his essay 'On the tendency of varieties to depart indefinitely from the original type'. Varieties were, he had noted in *Notebook 4* 'produced at birth the offspring differing from the parent. This offspring propagates its kind'. Furthermore, he believed that

---

*life told in an autobiographical chapter.* London: Murray, pp. 189–90; (3) Wallace, A.R. 1895. *Natural selection and tropical nature.* London: Macmillan, pp. 20–1; (4) Wallace, A.R. 1903. The dawn of a great discovery: 'My relations with Darwin in reference to the theory of natural selection'. *Black and White* 25: 78; (5) MLr: 360–3; (6) Wallace, A.R. 1908. Address [acceptance speech on receiving the Darwin-Wallace Medal on 1 July 1908], in *The Darwin-Wallace Celebration Held on Thursday, 1st July 1908, by the Linnean Society of London.* London: Longmans, Green & Co., pp. 5–11 [S656].

new varieties were constantly appearing—without reference to the environment. This means he envisioned tiger beetles, for example, constantly producing daughter varieties of varying colours. If those poorly matching the environment were destroyed by predators, only those that happened to match it or matched it best would survive the 'struggle for existence' famously outlined by Thomas Malthus. This differential survival of forms could explain how some beetles came to match their background so perfectly. If the mud or sand changed colour, for example, varieties of Wallace's tiger beetles could adapt by differential survival. The parent species however, still wearing the old colours, would die out, and thus be succeeded by the daughter variety that would in turn become or be classified as a new species. This variety could never return to the colour of the parent type since that was now a non-matching colour with the background.

After writing his essay, Wallace travelled to Gilolo around 18 February to explore and collect. A remark in his notebook referred to the poor collecting area on Gilolo shortly after his arrival but it was misdated '*Gilolo—Jan. 20th. 1858*'.[220] In that entry he mentioned again a struggle for existence. Once a plain in the tropics had been covered by grasses this would prevent the return of trees. Reminiscent of the language in his 2 March letter to Frederick Bates, Wallace noted that 'This for a long time puzzled me, but I think I have found the explanation'. He returned to Ternate on 1 March. The next mail steamer, always expected in the second week of the month, arrived on Tuesday 9 March. Most commentators have assumed that Wallace sent his essay to Darwin on this steamer, because it was dated February and Wallace later recollected sending it by 'the next post'.[221] The letter to Frederick Bates of 2 March was definitely sent on that steamer and its surviving postmarks reveal that it arrived in London on 3 June 1858—exactly as the mail schedules reveal it should have. Darwin, however, wrote to Charles Lyell that Wallace's Ternate essay arrived on 18 June. This discrepancy has long

---

[220] *Notebook 4*, p. 109.
[221] Wallace to A.B. Meyer, 22 November 1869, published in *Nature* 52 (1895): 415.

remained unexplained and has fuelled a small conspiracy industry by those eager to accuse Darwin of deceit and even plagiarism.

The March steamer did not bring the supplies Wallace still expected from Stevens, but it did bring another letter from Darwin (22 December 1857). From it, Wallace learned for the first time that the great Lyell was also impressed with the Sarawak Law paper. One of the few details known about Wallace's lost letter accompanying his Ternate essay is that it was a reply to this letter by Darwin, because Wallace referred to Lyell's interest in the Sarawak Law paper and asked Darwin to forward the Ternate essay to him. It is unlikely that Wallace could have responded on the same steamer. He is not known ever to have done so — perhaps the turnaround time was too brief. But more importantly, if we consider the postal route between Ternate and Darwin's home we find that the arrival date stated by Darwin, 18 June 1858, exactly matches the arrival of mail posted from Ternate on the next steamer, leaving in early April 1858.[222]

Having deposited his letter and essay at the Ternate post office, Wallace and four assistants journeyed on a trading schooner to the remote shores of Dorey, New Guinea. Apart from two German missionaries living on the nearby island of Mansinam, Wallace was apparently the only European resident on New Guinea at the time. He spent three and a half months at Dorey, beset by illness and constantly plagued by insects. Despite all his efforts, Wallace was never able to obtain good specimens of the Birds of Paradise, which after all had been his main reason for travelling to these remote shores. Tired and disappointed, Wallace and his team, minus one who died from dysentery, returned to Ternate on the same trading schooner on 15 August 1858. Here he wrote to Stevens about this unsuccessful journey, and sent off an article on the geography of the island to the Royal Geographical Society.

It was only on his return from New Guinea that Wallace learned of the historic events that were set off by his fateful letter to Darwin. He received letters from Darwin and Joseph Dalton Hooker with the news that Wallace's Ternate

---

[222]  Wyhe, J. van and Rookmaaker, K. 2012.

essay had been read together with manuscripts by Darwin before the Linnean Society on 1 July 1858. Sadly, neither letter survives. In recent decades this event has become a controversial story with many heated opinions about what was supposedly fair to Wallace.

Although Darwin's letter does not survive, Wallace copied an extraordinary extract from it on a blank page of *Notebook 4*. Darwin (13 July 1858) shared extensive details and a draft table of contents of his species book in progress. Although we can never know why this Darwin letter is lost, it is possible it was sent to someone else and hence Wallace made a copy of the table of contents to keep with him.

Wallace wrote to his mother and to Hooker on 6 October. He told his mother that he was well, had recovered from the New Guinea expedition, and was about to set off for the island of Batchian. He informed her of the exciting news. He was 'highly gratified' that Lyell and Hooker had 'immediately read [the Ternate essay] before the "Linnean Society". This insures me the acquaintance and assistance of these eminent men on my return home'. Our transcription preserves Wallace's original choice of words which shows that he considered calling Darwin and Hooker two of the 'greatest' naturalists in England but then settled on 'eminent', a term which he used twice.

Wallace's style in his letter to Hooker is very unusual. It was perhaps the most deferent letter Wallace ever wrote. This letter is often quoted, but what has been overlooked is that Hooker's letter (now lost) apparently informed Wallace of the reading of his essay but not its imminent publication. Presumably it was not deemed necessary to inform him once the essay had been made public. This seems to be the case because when writing to Stevens at the end of October Wallace was unsure if his paper was to be published or not. Wallace's frank remarks to his mother and later to Silk (30 November 1858), as well as the letter to Hooker, reveal unequivocally how delighted Wallace was that his essay had been read before the Linnean Society together with Darwin's earlier extracts.

# Ternate

## 41. To Frederick Bates, 2 March 1858

**Ternate**

*March 2nd. 1858*

My dear Mr. Bates

When I received your very acceptable letter (a month ago) I had just written one to your brother, which I thought I could not do better than send to you to forward to him, as I shall thereby be enabled to confine myself solely to the group you are studying & to other matters touched on in your letter. I had heard from Mr. Stevens some time ago that you had begun collecting exotic Geodephaga, but were confining yourself to one or two illustrations of each genus. I was sure however you would soon find this unsatisfactory.

Nature must be studied in detail, and it is the wonderful variety of the *species* of a group, their complicated relations & their endless modification of form, size & colour, which constitute the preeminent charm of the Entomologist's study. It is with the greatest satisfaction too, I hail your accession to the very limited number of collectors & students of exotic insects, & sincerely hope you may be sufficiently favoured by fortune, to enable you to form an extensive collection & to devote the necessary time to its study & ultimately to the preparation of a *complete & useful work*.[223] Though I cannot but be pleased that you are able to do so, I am certainly surprised to find that you indulge in the expensive luxury of from *3 to 7 specimens ! of a species*.[224] I should have thought that in such a very extensive group you would have found one or at most a pair quite sufficient. I fancy very few collectors of exotic insects do more than this, except where they can obtain additional specimens by gift or by exchange.

---

[223] Annotation by Frederick Bates: 'Don't I wish you may get it (F.B.)'.
[224] Annotation by Frederick Bates: 'A mistake'.

Your remarks on my collections are very interesting to me, especially as I have kept descriptions with many outline figures of my Malacca & Sarawak Geodephaga, so that with one or two exceptions I can recognize & perfectly remember every species you mention. I will first make a few remarks on your notes: *Cicindela elegans* [tiger beetle] is the *Megalomma elegans* of Westwood & I think it quite as well entitled to be separated as the *Heptadontas* which last seem to me not to differ from *Odontocheila*. Is *no. 61 Mac.* certainly *C. vigorsii*? It is certainly singular if so that I should not have found so widely a distributed a species before reaching Macassar. Is *60 Mac.* the true *C. heros*? & is the true *C. heros* Fab. very rare? I am inclined to think it is, because the little *Therates 59. Mac.* is undoubtedly *T. flavilabris* Fab. & *fasciatus* Fab. (varieties) & the localities for both are "Pacific Islands" brought home I think by *Labillardière*. In my 2nd. lot from Macassar I sent plenty of the *Therates*, but did not find *Heros* again. I took however 5 additional species of true *Cicindela* (one very handsome) making 11 species in all, a very large number for one locality.

I fear much I have lost the pretty genus *Collyris*. In Macassar was only 1 species & I have never seen it in any of the islands eastward. At Macassar I *once* saw a *Tricondyla* but the villain escaped up a tree, & I vainly searched for him for a month afterwards. I shall probably however meet with him when I visit the N. of Celebes where I expect lots of fine things. The Macassar *Collyris* seemed to me identical with *88. Sarawak*. The *Catadromus* you call *Boisduvalii*. Is it not the Javanese sp. & is not the name *C. tenebrioides*? I got but a solitary spec$^n$ in my second season at Macassar, but of the two rarest species of *Cicindela* which you have not yet, I have sent several specimens. Many of the insects you mention as desiderata from *Malacca* & *Sarawak* are *uniques* in my coll. Thus I believe 6 out of my 8 species of *Orthogonius* are so. It is the rarest genus I know of. The larger rare Malacca *Therates* was taken only at the foot of Mt. Ophir (5–6 specimens) & the pretty small species was taken only during 10 days collecting on my return to Singapore from Borneo (2–3 spec$^{ns}$).

It is in the highest degree improbable that I shall ever return to Sing. or Mal. again. Numbers of the best things I got in plenty, & the *Coleoptera* generally run so very small that another visit would not pay expenses. The smaller *tri-condyla* & *hepladonta* from Sarawak were also I think both unique, as probably also one in two of the *Collyris*. I wonder you have not noticed what I consider the *gem* of my Geodephaga, (889. Sar.) a *Catascopus* 10 lines long & very broad with a band of *rich purple* across the elytra shading in to the metallic green. I got only two examples one of M̶r̶ which goes to Mr. Saunders by agreement.

Soon after you receive this, my 2nd. Macassar collection ought to arrive, & I think it will well repay your examination. Some account of it will be in the Zoologist by this time I dare say.[225] I think it may be considered the most remarkable lot of *Carabidae* ever collected in the tropics in so short a time; almost all in *6 weeks* at the beginning of the rains, after a previous 6 weeks of the most terrible dearth of all coleoptera. I make out about 105 species of *Carabidae* of which 20 are minute things under 2 lines & many under a line! Are not some of these among the smallest Carabidae known? About 40 are truncatipennis, of which a dozen are lovely metallic or coloured species,—two or three very lovely. There is a very curious little thing allied to *Casnonia* with swollen thorax & long palpi, probably a new genus. Let me have your opinion of it. There is plenty of it sent.

The greater part of this collection will be probably new species. I spent hours daily on my knees in wet sand & rotten leaves, hunting the little things & picking up *Anthici* & *Pselaphidae* with the tip of my wetted finger. I shall be very much interested to have your remarks on the collection. Tell me if you think *any* or *how many* have been sent from Sing. or Sar. or are known from Java. Whether you think there are any new *genera*. Are the lot of little spotted species allied to our Bembidiidae? (365) is a lovely thing, unique, found in *foliage*! It is not much use your referring to the *numbers* here, as I had no time to take descriptive notes. They are put only for my notes of *station & habits*. You will often see two or three species with the same number. This is where they were

---

[225]   Wallace, A.R. 1858. [Letter dated 20 December 1857, Amboyna]. *Zoologist* 16 (191–2): 6120–4 [S44].

taken at the same time & place so that one note serves for all. My Aru collection was very poor in *Geodephaga*. Nothing remarkable but the *Therates labiata* & *Tricondyla aptera*, the two oldest known species but I believe not common in collections. I was in doubt if there are two sp. of the *Tricondyla*. One has much redder legs. They are not sexes as I took a pair "in cop." with similar legs. Try if you can find any other specific difference. They are found in the same places. I send you a pair in this letter. In Arru I did not see a *Cicindela*! In my small Amboyna coll. the Geodephaga are very few as I was too much occupied with the fine *Longicorns, Curculidae* & *Buprestidae* to search for the small ones.

Now with regard to your request for notes of habits, &c. I shall be most willing to comply with it to some extent, first informing you that I look forward to undertaking on my return to England a "*Coleoptera Malayana*,"[226] to contain descriptions of the known species of the *whole Archipelago*, with an essay on their geog. distribution, and an account of the habits of the genera & species from my own observations. Of course, therefore, I do not wish any part of my notes to be previously *published*, as this will be a distinctive feature of the work; so little being known of the habits, stations & mode of collecting exotic coleoptera.

As I have not much more room without making this a double letter, I will here tell you a little about the *Cicindelidae* only. The true Cicindelae vary considerably in habits. Some frequent almost exclusively sunny pathways, through open grounds & even public roads, such as 41, 42, 43 & 51 *Sing.* & 61, 62, 64, 65 *Mac.* Others are sea beach insects as the *C. tenuipes* & the *Baly species*—the former singularly agreeing in colour with the white sand of Sarawak, the latter with the dark volcanic sand of its habitat. Others prefer

---

226   Wallace later assisted in the publication of Pascoe, F.P. 1864–1869. Longicornia Malayana; or, a descriptive catalogue of the species of the three longicorn families Lamiidæ, Cerambycidae, and Prionidae, collected by Mr. A. R. Wallace in the Malay Archipelago. *Transactions of the Entomological Society of London* (series 3) 3: 1–712, 24 pls.

Fig. 17. The 2 March 1858 letter to Bates

river banks. The two Lombock sp. were found always a little way inland on the same coloured dark sand, but I never found them on the sea beach, so also 63 & 126 Macassar, frequent river banks on sand of a lighter col[r] than that of Baly & Lomb. but darker than that of Sarawak, *as are the insects*. Another n.s. in the last Mac. coll. was found in the soft shiny mud of salt creeks, with which its colour so exactly agrees that it was ~~exact~~ perfectly invisible except for its shadow!

Such facts as these puzzled me for a long time, but I have lately worked out a theory which accounts for them naturally. The rule however is by no means without exception. The bright coloured species are visible enough wherever they are. C. *Heros* frequents shady path in the woods settling on the ground or on the foliage & flying slowly to short distances with a distinct buzzing noise. Another sp. sent in 2nd. coll. (313 Mac) frequents similar situations but flies much quicker; its bright golden buff spots render it very conspicuous, & it emits when captured a fine rosy odour like the *Aronias*, which I have not observed in any other species of the restricted genus *Cicindela*. I see I must defer my notes on the other genera to my next letter as I want to say something about other matters.

You appear to consider the state of Entomological literature flourishing & satisfactory;—to me it seems quite the contrary. The number of *unfinished* works & of others with *false titles* are disgraceful to science. Dejean's *Species General*[227] was meant to be finished but is not. Burmeister's Handbook of *Entomology*[228] could never have been meant to be finished on the scale it is begun on—it has a false title. "*Annulosa* Javanica" contains a part of the Coleoptera only. G.R. Gray's grand title of "*Entomology* of Australia"[229] dwindles to the smallest family (Phasmidae) of the smallest order of Insects. Mr. Woollaston[230] never intends to complete an *Insecta* Maderensea. Why then should he give it that grand title? There exists not one *completed* work on any extensive group of Coleoptera published within the last 20 years, except those of one man (M. Lacordaire).[231] There exists not a coleopterous fauna of any one tropical district, of any one extra european country! And greatest disgrace of all, there exists not any work on the Coleoptera of Europe! (complete). Is this satisfactory?

227  Dejean, P.F.M.A. 1825–1838. *Species général des coléoptères de la collection de M. le comte Dejean.* Paris: Méquignon-Marvis, 6 vols.

228  Burmeister, H. 1832–1855. *Handbuch der Entomologie.* Berlin, 5 vols.

229  Gray, G.R. 1833. *The entomology of Australia, in a series of monographs. Part I: Monograph of the genus Phasma.* London.

230  Thomas Vernon Wollaston (1822–1878), entomologist and author of *Insecta Maderensia; being an account of the insects of the islands of the Madeiran group.* London: John Van Voorst, 1854.

231  Lacordaire, T. 1854–1876. *Histoire naturelle des insectes: genera des coléoptères.* Paris, 10 vols.

Does there exist a satisfactory modern local coleopterous fauna of any one country in Europe? Mulsant's is I believe not yet finished.[232] There is I understand one of *Hungary* but with very brief analytical characters only to the species. What a shame that the Entomologists of Germany & France with such a large proportion of the species at their doors have not yet produced a "Coleoptera Europae"!! Is there a *Catalogue* of the Coleoptera in any one of the National Museums of Europe—or *any hope of one?* If you go to America & ~~wha~~ want to know anything of the U.S. Coleoptera. what have you to guide you but scattered papers in periodicals? All this is to me *very unsatisfactory.* Lacordaire's *Genera* deserves all praise, but it would have been much more satisfactory had he kept back the first volumes to have published all together, *complete up to a given date.* I only hope he may finish it. I trust the plates may be good, but not expensive. I admire those of *Mulsant,* I have his Longicorns here. Monographs of Col[a] with coloured plates are luxuries. The *Cicindelidae* may be done. The *Carabidae* cannot be. I forgot the *Cetoniidae* Gory's monograph[233] but they are all large & showy species & not very numerous.

In *Geodephaga* no doubt the Amazon is superior to the E. Archipelago;—yet if the Ega collections were entirely left out that superiority would be by no means so apparent & I believe the richest islands for Geodephaga are Java & Sumatra, wh. I have not visited.[234] Bowring has a true *Carabus* from Java! Again considering the vast extent of the Amazon region, its position in the centre of a vast continent, its streams converging from the Andes for a distance of more than a thousand miles & its excessive productions in all departments of nature, & the wonder will be, not that these islands are inferior in one group, but that they are not vastly inferior in all. The whole Archipelago could be set down in the forest plains of the Amazon, the separate islands could be hidden and lost in it! Again, I find that in journeys, sickness, & time necessarily spent in towns far from the

[232]    Mulsant, M.E. 1839–1868. *Histoire naturelle des coléoptères de France.* Paris: Masson, 10 vols.

[233]    Gory, H.L. 1833–1836. *Monographie des Cétoines et genres voisins.* Paris.

[234]    Annotation by Frederick Bates: 'I am afraid Mr. W. [Wallace] requires too much at the present time – He evidently has yet to learn of the *difficulties* (F.B.)'.

forests, I have lost one whole year out of the 3¾ I have been in the East.—Yet my total species both in Coleoptera & in all Insecta compares very favourably with your brother's of a much longer period, and moreover as an experienced & persevering collector of Coleoptera he is decidedly my superior.

I think therefore on the whole we may say that the Archipelago is *very rich*, & will bear a comparison even with the richest part of S. America. In the country between Ega & Peru there is work for *50 collectors* for *50 years*. There are hundreds & thousands of Andean valleys every one of which would bear exploring. Here it is the same with islands. I could spend 20 years here were life long enough, but feel I cannot stand it, away from home & books & collections & comforts, more than four or five, & then I shall have work to do for the rest of my life. What would be the use of accumulating materials which one could not have time to work up? I trust your brother may give us a grand and complete work on the Coleoptera of the *Amazon Valley*, if not of all S. America.[235] My paper is full so I must now conclude with best wishes.

Yours faithfully

Alfred R Wallace

F. Bates Esq.

What is the gist of Wollaston's book on varieties &c.[236] From some extracts from his "Insecta Maderensia" I have seen, I do not think much of his philosophizing powers. What on Earth is Gosse's "*Omphalos*" & what is the New Law he has discovered?[237]

*viâ Southampton.*[238]

---

[235] Annotation by Frederick Bates: 'x A mere *trifle*, that would be for the extreme leisure hours. – X Why not "go the whole hog" at once !!! (F.B.)'.

[236] Wollaston, T.V. 1856. *On the variation of species with special reference to the Insecta; followed by an inquiry into the nature of genera.* London.

[237] Philip Henry Gosse (1810–1888) published *Omphalos: an attempt to untie the geological knot.* London: J. van Voorst, 1857.

[238] There are three postmarks around the address, plus possibly a fourth partly printed and hardly visible. The first postmark reads: SINGAPORE P.O. 21 APR 58; the second: LONDON BC JU 3 58; the third: LEICESTER C JU 3 58.

Fred<sup>r</sup>. Bates Esq.

5, Napier Terrace

Aylestone Road

Leicester

*England.*

*ARW.*

## 42. From James Motley, 22 May 1858

### Kalangan Banjermassing

*May 22, 1858*

Alfred R. Wallace Esq.

My dear Sir,

I have received only a few days ago your letter from Ternate dated Jan 4th. I have naturally heard of your wanderings in these countries, & I have long hoped to meet with you, a good fortune to which it appears I am not destined. I believe indeed that it is very unfortunate that you have not visited Banjermassing. There is no spot in Borneo which allows such easy access to the interior, from its enormous rivers & the very friendly character of the Dyak tribes & though I have only within two or three months commenced collecting them, I have reason to suppose that the country is exceedingly rich in insects. I have dispatched to M<sup>r</sup> Dillwyn[239] a collection in other branches of natural history, among which are some rare if not new species. It consists of between 500 & 600

---

[239] Lewis Llewelyn Dillwyn (1814–1892), industrialist and naturalist. He was a friend of Motley and together they published a book on the natural history of Labuan (1855). On Motley's life and work, see Bastin, J. 1987. James Motley and his contributions to the natural History of Labuan. *Journal of the Malaysian Branch of the Royal Asiatic Society* 60 (2): 43–54; and Walker, A.R. 2005. James Motley (1822–1859). *Minerva, Transactions of the Royal Institution of South Wales* 13: 20–37.

specimens comprising 185 species of birds, 35 mammalia & about 50 amphibia, the generality of which are very beautiful specimens, but my time is so fully employed here, that I have collected but few of these myself.

I have in my service a *[word destroyed]* collector, but I have not yet brought him into the way of collecting insects. He is however a clever & enterprising fellow, afraid of nothing, & well acquainted with the interior, & should it fall in your way to come here, I shall with pleasure send him with you into the heart of Borneo, & I really believe we should both profit by such an arrangement.

I have lately commenced a collection of Coleoptera, Hemiptera &c &c in conjunction with one of my mining pupils here Mr van Heekeren,[240] this collection is destined for Leyden, but we shall doubtless have many duplicates which I shall be most happy to exchange with you, however you must not expect to receive them very soon, as we have very little time, & have been at work about 2 months only. I consider that we go on pretty well. For the Lepidoptera we have also good intentions, but we wait for more boxes, & as you by this time very well know, this is a horribly slow country. We are also terribly at a loss of pins, & if you could give me a hint about the best way of procuring them & the terms in which to order them, as to numbers or sizes, you would do me a great favour.

Among the specimens sent to M<sup>r</sup> Dillwyn are two sculls & one skin of the great orang Hutan of very uncommonly large size. I am pretty sure such specimens have not been seen in Europe. The animal from which the skin was taken measured 11 feet over the outstretched anus, unhappily he received a ball in the mouth & the canines on one side are shattered.[241] While in the far East should you hear any tidings of an animals which seems somewhat apocryphal, I should be very much obliged for them. I mean a species of alligator of which I have had reports from the native traders to the Arru Islands & New Guinea, they call it buaya Kodok or toad alligator & it is reported to have a tail less than

---

[240] Jacobus Johannes van Heekeren (1834–1859) was (together with Motley) present at the official opening of the Julia Hermina mine near Banjermassin in June 1856 (*Javabode*, 26 July 1856).

[241] Annotation by Wallace: 'my largest 7.8!'

half the length of the body. I am much interested in the saurians, & of this species either a young one in spirits or an adult scull would be exceedingly acceptable, if indeed the species exists. A Saurian with a carapace or a tortoise with a sauroid tail & head, whichever it may be, is also said to exist in New Guinea.

I am not altogether without hopes of seeing you in this part of the world, & hoping to have the pleasure of an occasional letter from you,

I remain, my dear Sir,

Yours,

James Motley.

P.S. I omitted to tell you that we have here an acquaintance of yours, from whom I have frequently heard of you, Mr van der Tak[242] of the Handel Maatschappij. He is well I believe but stationed too far from me to be asked if he has any message for you.

Alfred R. Wallace Esq.

Mes[srs] Duivenboden

Ternate

James Motley[243]

---

## 43. From Charles Darwin, [13 July 1858][244]

### Sketch of Mr. Darwin's "Natural Selection"

Chap. I. On variation of animals & plants under domestication, treated generally.

" II. do. do. treated specifically, external & internal structure of Pigeons & history of changes in them.

---

[242] The Nederlandsche Handel Maatschappij (Netherlands Trading Society) opened a branch in Batavia in 1826 and in Singapore in 1858. It was responsible for trading and shipping for the government. Van der Tak is unidentified.

[243] There is a postmark: Ongefrankeerd. Zee Brief. Banjarmassing. 'Zee Brief' is 'letter by sea'. 'Ongefrankeerd' means that the recipient paid for the postage.

[244] Extract of Darwin's letter written by Wallace in *Notebook 4*.

" III. On intercrossing, principally founded on original observations on plants.

" IV. Varieties under Nature.

" V. Struggle for existence, malthusian doctrine, rate of increase,—checks to increase &c.

" VI. "Natural Selection" manner of its working.

" VII. Laws of variation. Use & disuse reversion to ancestral type &c. &c.

" VIII. Difficulties in theory. Gradation of Characters.

" IX. Hybridity.

" X. Instinct.

" XI. Palaeontology & Geology.

" XII. and XIII. Geog. distribution.

" XIV. Classification, Affinities, Embryology.

## 44. To Samuel Stevens, 2 September 1858

**Ternate**

*September 2, 1858*

When I arrived here from New Guinea, about a fortnight ago, I found your two letters of January and March, noting the safe arrival of the Aru collections and the advantageous disposal of the birds: they gave me the greatest pleasure and satisfaction, and the interest the collections appear to have excited was a great encouragement to me; and I assure you I stood in need of some encouragement, for never have I made a voyage so disagreeable, expensive and unsatisfactory as the one now completed.

I suffered greatly from illness and bad or insufficient food, and am only now just sufficiently recovered to work hard at cleaning and packing my collections: my servants suffered as much as myself; two or three were always sick, and one of my hunters died of dysentery. My collections will greatly disappoint you and my other friends,—more than they do myself,—because you will be expecting something superior to Aru, whereas they are very inferior in fine things. First and foremost, all my hopes of getting the rare paradise birds have vanished, for not only could I get none myself, but could not even purchase a single native skin! and that in Dorey, where Lesson[245] purchased abundance of almost all the species: he must have been there at a lucky time, when there was an accumulated stock, and I at a most unpropitious one.

It is certain, however, that all but the two common yellow species are very rare, even in the places where the natives get them, for you may see hundreds of the common species to perhaps *one* of either of the rarer sorts. There are some eight or ten places where most of the birds are got, and from each I doubt if there is more on the average than one specimen per annum of any other than the Paradisea papuana; so that a person might be several years in the country, and yet not get half the species even from the natives: yet they are all common in Europe!

I sent two of my servants with seven natives a voyage of one hundred miles to the most celebrated place for birds (Amberbaki, mentioned by Lesson), and after twenty days they brought me back nothing but two specimens of P. papuana and one of P. regia: they went two days' journey into the interior without reaching the place where the birds are actually obtained; this was reported to be much further, over two more ranges of mountains.[246] The skins pass from village to village till they reach the coast, where the Dorey men buy them and sell to any trading vessels. Not one of the birds Lesson bought at Dorey was killed there; they came from a circuit of two hundred or three hundred miles.

My only hope lies now in Waigiou, where I shall probably go next year, and try for P. rubra and P. superba. Even of P. papuana I have not got many, as my boys

---

245  Lesson, R.P. 1826–1830. *Voyage autour du monde*. Paris.

246  On 27 May 1858 Wallace sent his boys by boat to Amberbaki, a village about 100 miles to the west, to shoot, find, or buy Birds of Paradise. Local people also made this journey to buy supplies and items for trade, see Wallace, A.R. 1862. On the trade of the Eastern Archipelago with New Guinea and its islands. *Journal of the Royal Geographical Society* 32: 127–37 [S65].

had to shoot them all themselves; I got nothing from the natives at Dorey. You will ask why I did not try somewhere else when I found Dorey so bad: the simple answer is, that on the whole mainland of New Guinea there is no other place where my life would be safe a week: it is a horribly wild country; you have no idea of the difficulties in the way of a single person doing anything in it. There are a few good birds at Dorey, but full half the species are the same as at Aru, and there is much less variety! My best things are some new and rare lories.

In insects, again, you will be astonished at the mingled poverty and riches: butterflies are very scarce; scarcely any Lycaenidæ or Pieridæ, and most of the larger things the same as at Aru. Of the Ornithoptera I could not get a single male at Dorey, and only two or three females; I got two from Amberbaki and two from the south coast of New Guinea, from the Dutch exploring ship.[247] Of Coleoptera I have taken twice as many species as at Aru; in fact, I have never got so many species in the same time; yet there is hardly anything fine: no Lomopteræ,—in fact, not one duplicate Cetonia of any kind, and only two solitary specimens of common small species! No Lucani! perhaps nowhere in the world are Lamellicornes so scarce,—only fourteen out of 1040 Coleoptera, and most of them small and unique specimens.

Of Longicornes there are full as many as at Aru; many the same, but a good number of new and interesting species. Curculionidæ very rich; some remarkable things, and the beautiful Eupholus Schœnherri and E. Cuvieri; the former rather abundant. There is a very pretty lot of Cicindelidæ; two Cicindelas and three Therates will probably be new to the English collections; they are C. funerata, *Bois.*, a very pretty species, with a peculiar aspect; C. d'Urvillei; also a small new species, near C. funerata, very scarce. Therates basalis, *Dej.*, a very pretty species, I have sent

---

[247] A Dutch-built iron paddle steamer, the *Etna*, arrived at Dorey with members of a Dutch Commission in May 1858. Nederlandsch Indische Commissie. 1862. *Nieuw Guinea: ethnographisch en natuurkundig onderzocht en beschreven in 1858 door een Nederlandsch Indische commissie*. Amsterdam: F. Muller, p. 65. Wallace seemed displeased with their presence as they were given first choice in obtaining any natural history specimens. Some members of the Commission suggested that Wallace might be a spy for his country.

a good many of; T. festiva, *Dup.* (I think), a pretty brilliant little species, not common, and another of the same size, and, I think, quite new, rufous and black marked, also scarce; T. labiata and Tricondyla aptera are the same as sent from Aru.

I have never before found so many species of Therates in one place: they form quite a feature in the Entomology of Dorey. Carabidæ [ground beetles] were very scarce: I picked up, however, some pretty things, especially two most brilliant Catascopi, but both unique. For a long time I took no Staphylinidæ [rove beetles]: at last I found a station for them, and by working it assiduously I got between eighty or ninety species: some are the handsomest of the group I have yet taken, and there are many curious and interesting forms. Talk about Brachelytra being rare in the tropics! of their place being supplied by ants, &c., &c.! why, they are absolutely far more abundant in the tropics than anywhere else, and I believe also more abundant in proportion to the other families.

I see in the 'Zoologist' two local lists of Coleoptera (Dublin and Alverstoke),[248] in which the numbers of Staphylini are 103 and 106 species respectively; these are the results of many years collecting by several persons, and in a country where all the haunts and habits of the tribe are known; here, in two localities (Macassar and Dorey), I have taken at each nearly the same number of species, in three months' collecting, on a chance discovery of one or two stations for them, and while fully occupied with extensive collections of all orders of insects, in a country where every other one is new. The fair inference is, that in either of these localities Staphylini are really ten times as numerous as in England; and there is reason to believe that any place in the tropics will give the same results, since in the little rocky island of Hong-Kong Mr. Bowring has found nearly 100 species; yet Dr. Horsfield,[249] who is said to have collected assiduously in Java, did not get a solitary species.

[248] Hogan, A.R. 1853–1854. Catalogue of Coleoptera found in the neighbourhood of Dublin. *Zoologist* 11: 4134–6; and *Zoologist* 12: 4195–9, 4338–41. Adams, A. and Baikie, W.B. 1857. A systematic list of Coleoptera found in the vicinity of Alverstoke, South Hants. *Zoologist* 15: 5545–55.

[249] Thomas Horsfield (1773–1859), an American naturalist who travelled and collected in Java and Sumatra between 1801 and 1819. He published *Zoological researches in Java and the neighbouring islands* in 1824.

My next richest and most interesting group is that of the Cleridæ, of which I have about fifty species, perhaps more, for they are very puzzling: I have never got so many in one locality, nor should I now had I not carefully set them out and studied their specific characters, and thus separated many which would otherwise certainly have escaped notice. In another small and obscure group, the Bostrichidæ [auger beetles] and allied Scolytidæ, I obtained no fewer than thirty-eight species, whilst the Lampyridæ [fireflies] and allied groups were in endless and most puzzling variety. I have also got an exceedingly rich and interesting series of Galerucidæ and Chrysomelidæ.

The Elaters are small and little interesting. The Buprestidæ [metallic wood-boring beetles] also are very inferior, and of the only fine species (*Chrysodema Lotinii*) I could only obtain a single pair. With so many minute Coleoptera I could not give much attention to the other orders; there are, however, some singular Orthoptera, and among the Diptera a most extraordinary new genus, the males of which are horned; I have three species, in two of which the horns are dilated and coloured, in the other long, slender and branched; I think this will prove one of the most interesting things in my collection. One would have thought Dorey would have been just the place for land shells, but none were to be found, and the natives hardly seemed aware of the existence of such things; I have not half-a-dozen specimens in all.

Although Dorey is a miserable locality,—the low ground is all mud and swamp, the hill very steep and rugged, and there are only one or two small overgrown paths for a short distance, my excursions were almost entirely confined to an area of about a square mile,—yet the riches in species of Coleoptera, and a considerable number of fine remarkable forms of which I could obtain only unique examples, sufficiently show what a glorious country New Guinea would prove if we could visit the interior, or even collect at some good localities near the Coast.

You ask me if I go out to collect at night; certainly not, and I am pretty sure nothing could be got by it: many insects certainly fly at night, but that is the reason why they are best caught in the day in their haunts, or else by being attracted to a light in the house.[250] Besides a man who works, with hardly half an hour's intermission, from 6 A.M. till 6 P.M., four or five of the hottest hours being spent entirely out of doors, is very glad to spend his evenings with a book (if he has one) and a cup of coffee, and be in bed soon after 8 o'clock. Night work may be very well for amateurs, but not for the man who works twelve hours every day at his collection.

I am perfectly astonished at not yet meeting with a single Paussus [ground beetles]; Several are known from the Archipelago, and have been taken in houses and at light, yet my four years look-out has not produced one. How very scarce they must be! You and Dr. Gray[251] seem to imagine that I neglect the mammals, or I should send more specimens, but you do not know how difficult it is to get them: at Dorey I could not get a single specimen, though the curious tree-kangaroos are found there, but very rare: the only animal ever seen by us was the wild pig.

The Dutch surveying steamer bought two kangaroos at Dorey whilst I was there: it lay there a month waiting for coal, and during that time I could get nothing, everything being taken to the steamer. I send from Dorey a number of females and young males of Paradisea papuana; these females have been hitherto erroneously ascribed to P. apoda, of which I am now convinced my specimen from Aru is an adult female; it is totally brown: the females of P. papuana are smaller than the young males, and have the under parts of a less pure white: the bird figured by Levaillant[252] as the female of P. papuana is a male of the second year which has acquired the green throat in front, but not the long feathers of the tail or flanks: to all the female specimens I have attached tickets,—all not ticketed are males.

---

[250]  Stevens was a pioneer of night collecting and encouraged others to do so. See Douglas, J.W. 1856. *The world of insects*. London, p. 203.

[251]  John Edward Gray (1800–1875), keeper of zoology at the British Museum in London, 1840–1874. Gunther, A.E. 1975. *A century of zoology at the British Museum through the lives of two keepers 1815–1914*. London: Dawson.

[252]  François Levaillant (1753–1824), a French ornithologist who wrote one of the grandest illustrated books on birds of paradise: *Histoire naturelle des oiseaux de paradis et des rollier*. Paris, 1801–1806. On Levaillant's collections and discoveries, see Rookmaaker, L.C., Mundy, P.J., Glenn, I., and Spary, E.C. 2004. *François Levaillant and the birds of Africa*. Johannesburg: Brenthurst Press.

Whilst the Dutch steamer was at Dorey a native prow came from the Island of Jobie, and bought two specimens of Atrapia nigra, which were sold to a German gentleman,[253] who is an ornithologist, before I knew any thing of them: I believe that island is their only locality, and the natives are there very bad, treacherous and savage. That is also the country of the rare species of crown pigeon (*Goura Victoriæ*); a living specimen of this was also purchased on board the steamer. I have great thoughts, notwithstanding my horror of boat work at sea (for a burnt child dreads the fire) and my vow never to buy a boat again, of getting up a small craft and thoroughly exploring the coasts and islands of the Northern Moluccas, and to Waigiou, &c.; it is the only way of visiting many most interesting places,—the Eastern coast of the four peninsulas of Gilolo, the Island of Guebe, half-way between Gilolo and Waigiou, a most interesting spot, as Gilolo and Waigiou possess quite distinct Faunas.

A.R. Wallace

---

## 45. To Henry Norton Shaw, September 1858

**Ternate**

*Sept. 1858*

My dear D^r Shaw —

I send you a few notes ab^t. New Guinea to be read at the Royal Geographical Society's meeting if you think proper.[254] Afterwards will you have any objection

---

[253]  Carl Benjamin Hermann von Rosenberg (1817–1888), German naturalist working as draughtsman for the Dutch surveying expedition. Rosenberg's account of his travels between 1839 and 1871 was published in 1878 as *Der Malayische Archipel*, Leipzig: G. Weigel.

[254]  Wallace's article was read and discussed at the Royal Geographical Society, 14th meeting of 27 June 1859, with the Earl of Ripon, president, in the chair. An abstract and a review of the discussions are found in the printed *Proceedings of the Royal Geographical Society of London* 3 no. 6 (late 1859): 358–61, and the full text followed in 1860 in the *Journal of the Royal Geographical Society* 30: 172–7 [S51].

to send it to the Athenaeum for publication as it contains nothing worthy of insertion in the Society's Transactions.

With best wishes,

I remain,

Yours very faithfully,

Alfred R. Wallace.

D$^r$ *Norton Shaw*

## 46. To Frances and Thomas Sims, 6 September 1858

**Ternate**

*Sept. 6th. 1858*

My dear Fanny

In the box which will probably reach Mr Stevens two months after this, is one of Dumas' best romances "La Reine Margot" wh. was given me by a friend. It is a wonderful story about the massacre of Bartholomew &c. &c. &c. & will help you to pass some long evenings or solitary hours.

I hope you will not think of going back to Conduit St. now you have once left it. Thomas' country connexion is likely to increase & be more certain & more profitable than the portrait work, beside that it is so much less harrassing. I hope G.C.S. [George Silk] *will* get married at last. He has written me a lot about the ladies in question. Let me know how the affair goes on when you write.

With best love, I remain

Dear Fanny, Your affectionate Brother,

Alfred R. Wallace.

*Mrs. Sims*

Dear Thomas

I have just seen in the Athenaeum an account of Claudet's great discovery of the *Stereomonoscope* though I cannot clearly see *why* it sh^d ~~be~~ produce the effects described.[255] Of course you have seen it. Is the effect as good & deceptive as in the common stereoscope.—What a grand thing to throw two magnified pictures on a transparent screen & show views & figures of the size & solidity of life! to a large audience. Can it be done? I sh^d think the spectator must be within a certain limited range both of distance & direction to get the proper effect.

The photographic *compositions* by Rejlander puzzle me.[256] How does he combine his negatives without showing the *joints*, or are they touched up afterwards to cover them. Again the Photographic Engravings are all new to me. Are they really good?

Now I have to recommend you a periodical "The Family Herald" only 1^d per week.[257] You probably know it & perhaps as I did at first, *despise* it. I took however an odd volume to New Guinea with me & of course was obliged to read it every line, & what was my astonishment to find in the leading articles a series of essays on the greatest questions of the day, which for *wit*, true *philosophy* & *novelty* are in my opinion unsurpassed in our whole periodical literature.

Besides there is a host of useful matter & lots of amusing & clever tales for odd half hours. Do take it, & when I return I will buy the back volumes wh. are double price, for the sake of the leading articles. Have you been to hear Spurgeon?[258]—the

---

[255]   Antoine François Jean Claudet (1797–1867). Claudet, A. 1857. On the stereomonoscope, a new instrument by which an apparently single picture produces the stereoscopic illusion. *Proceedings of the Royal Society of London* 9: 194–6.

[256]   Oscar Gustave Rejlander (1813–1875), a Swedish photographer in London who in 1857 made *The Two Ways of Life*, a seamlessly montaged combination print made of 32 images. Rejlander collaborated with Darwin for his book *The expression of the emotions* (1872), London: John Murray.

[257]   *The Family Herald: A domestic magazine of useful information & amusement* (1843–1940) was a weekly story paper established by James Elishama Smith.

[258]   Charles Haddon Spurgeon (1834–1892), pastor of the Particular Baptist Church in London and well-known as an author and preacher, often speaking to audiences numbering in the thousands.

*Athenaeum* cuts him up as a regular ranter. Is it your friend Mr. Fry who is publishing the Photographs of living celebrities?[259] Can't you get a set to copy?

Yours affectionately,

Alfred R. Wallace.

Thos. Sims Esq.

————

## To [Samuel Stevens], 5 October 1858

*Notes for Mr. Foxcroft from AR. Wallace*[260]

To collect Insects successfully in the Tropics you must get close to or in the *true virgin forest.*—If possible get a patch of forest cleared & a hut put up to live in—you will then get *double* the collections in the same time. On the trees cut down & drying near your house you will get hosts of *Longicorns* & *Curculios.*—As the bark rots, it will swarm with *Carabi* & *Staphilinidae*—On the bark of fallen trees in the sunshine *Buprestidae*, at night good things will come to a lamp, beat *dead leaves* assiduously, they produce good things especially when thick & damp—Rare Butterflies & variety of species can only be found in the forest paths—The *very finest* ground is where *new* roads or extensive clearings are being made through lofty forest—Seek out for such if possible & you will be well rewarded.—If none such can be found, at all events have few large trees cut down near

---

259   The *Photographic portraits of living celebrities* by Henry Maull (1829–1914), with text by Herbert Fry and Edward Walford, included 40 pictures issued from 1856 to 1859.

260   James Foxcroft (d. 1861), a commercial entomological supplier and collector. He advertised for subscribers to invest in his collecting expeditions to receive a share of the resulting collection. Foxcroft set out for Sierra Leone in 1858.

your house—They will produce good things morning & evening & at spare times when you cannot go far, palm tree flowers & the sap of the toddy palm, attract *Cetonias* & *Lucani.*—Living trees produce nothing except larvae, it is useless to search on them.—Under stones & on sandy places & on bare ground, little or *nothing*, the forest is the place for every thing fine & *swampy* forest is generally the best especially on the slopes of mountains. — From 9 am to 2 pm I find the best general collecting time—you would much oblige me by registering your captures daily & would find it interesting thus

| June 1 | *Coleoptera* | *Lep^d* | *Other orders* |
|---|---|---|---|
| | 35 species | 18 species | 44 species |

and at the end of each month make a summary of the N° of species collected in each family & order—Insects are seldom so abundant in the tropics as in England except just about & *after* the wet season. Four or five hours a day must be spent in diligent search & you will then be astonished at the new & fine things constantly turning up.

Hoping these notes the result of 8 years tropical collecting may be of use —

yours truly, Alfred R. Wallace

Ternate Moluccas. Oct 5th 1858

Direct all cases if sent by mail steamer

M^r. S. Stevens.

24 Bloomsbury Street.

London

Care of

Mr. J. Deal Jun.

Custom House Agent

High Street

Southampton

*with great care*

Direction for Collecting in the Tropics by A.R. Wallace
for M[r] H. Squires[261]

## 48. To Mary Ann Wallace, 6 October 1858

*October 6, 1858*

I was much amused with Fanny's interview with D[r] [John Edward] Gray and
Prof: [Richard] Owen at the Museum. The *latter* is a delightfully pleasing and
polite man, D[r] Gray rather *surly*, but *no doubt* can be polite enough *to Ladies*.
He has been on bad terms with me, because when in England I made some
remarks about on the museum which he did not like, but since my recent collecting he has bought of them *very liberally* and has named one of my new
animals after me, which I look upon as an "olive-branch" of reconciliation.—I
fear the Museum is too gloomy for Thomas Sims to take any good views in.

I sh[d] like a copy of the Paradise birds when printed, but nothing I have so far
received *yet* has come in good condition, the delicate lights fade into white
patches.

I have just returned from a short trip, & am now about to start on a longer
one, but to a place where there are some soldiers, a Doctor & Engineer[262] who
speaks English, so if it is good for collecting I shall stay there some months. It
is "Batchian", an island on the S.W. side of "Gilolo", about 3 or 4 days' sail from
Ternate. I am now quite recovered from my New Guinea voyage and am in
good health.

---

[261]  Henry Squire (d. before 1861), insect collector in England, Scotland, and Rio
de Janeiro.

[262]  Otto Fredrik Ulrich Jacobus Huguenin (1827–1871) joined the Dutch East
Indies government service in 1850 and by 1859 was head of the Mining Service
to oversee the mine work on Batchian.

I have received letters from Mr Darwin & Dr Hooker two of the ~~greatest~~ most eminent Naturalists in England which has highly gratified me. I sent Mr Darwin an essay on a subject in which he is now writing a great work. He shewed it to Dr Hooker & ~~Mr Darwin~~ Sir C Lyell, who thought so highly of it that they immediately read it before the 'Linean Society'. This insures me the acquaintance and assistance of these eminent men on my return home.

M^r. Stevens also tells me of the great success of the "Aru" collection, of which £1000 worth has actually been sold. This makes me hope I may soon realise enough to live upon and carry out my long cherished plans of a country life in old England.

If I had sent the large & handsome shells from Aru, which are what you expected to see, they would not have paid expenses, whereas the cigar box *full* of *small ones* has sold for £50. You must not think I shall always do so well as at Aru - perhaps never again, because no other collections will have the *novelty*, all the neighbouring countries producing birds & Insects very similar, and many even the very same. Still, if I have health I fear not to do very well. I feel little inclined now to go to California; as soon as I have finished my exploration of this region I shall be glad to return home as quickly and cheaply as possible. It will certainly be by way of the Cape or by 2^nd class overland.

May I meet you, dear old Mother, and all my other relatives and friends, in good health. Perhaps John & his Trio will have had the start of me.

## 49. To Joseph Dalton Hooker, 6 October 1858

**Ternate, Moluccas**

*Oct. 6. 1858.*

My dear Sir

I beg leave to acknowledge the receipt of your letter of July last, sent me by Mr. Darwin, & informing me of the steps you had taken with reference to a

Fig. 18. Joseph Dalton Hooker

paper I had communicated to that gentleman. Allow me in the first place sincerely to thank yourself & Sir Charles Lyell for your kind offices on this occasion, & to assure you of the gratification afforded me both by the course you have pursued & the favourable opinions of my essay which you have so kindly expressed. I cannot but consider myself a favoured party in this matter, because it has hitherto been too much the practice in cases of this sort to impute *all* the merit to the first discoverer of a new fact or a new theory, & little or none to any other party who may, quite independently, have arrived at the same result a few years or a few hours later.

I also look upon it as a most fortunate circumstance that I had a short time ago commenced a correspondence with Mr. Darwin on the subject of "Varieties", since it has led to the earlier publication of a portion of his researches & has secured to him a claim to priority which an independent publication either by myself or some other party might have injuriously effected;—for it is evident that the time has now arrived when these & similar views [exist] *will* be promulgated & *must* be fairly discussed.

It would have caused me much pain & regret had Mr. Darwin's excess of generosity led him to make public my paper unaccompanied by his own much earlier & I doubt not much more complete views on the same subject, & I must again thank you for the course you have adopted, which while strictly just to both parties, is so favourable to myself.

Being on the eve of a fresh journey I can now add no more than to thank you for your kind advice as to a speedy return to England;—but I dare say you well know & feel, that to induce a Naturalist to quit his researches at their most interesting point requires some more cogent argument than the prospective loss of health.

I remain

My dear Sir

Yours very sincerely

Alfred R. Wallace

*J. D. Hooker, M.D.*

Jos. D. Hooker, M.D. F.R.S.

# BATCHIAN AND TERNATE
## 9 October 1858–9 June 1859

After his return from New Guinea, Wallace decided to explore the other spice islands strung along the west coast of Gilolo. All were practically unknown to naturalists. A few days after returning to Ternate he set out to the large island of Batchian, south of Ternate. His team included 'my Bornean lad Ali, who was now very useful to me; Lahagi, a native of Ternate, a very good steady man, and a fair shooter, who had been with me to New Guinea; Lahi, a native of Gilolo, who could speak Malay, as woodcutter and general assistant; and Garo, a boy who was to act as cook' besides 'Latchi, as pilot. He was a Papuan slave, a tall, strong black fellow, but very civil and careful. The boat I had hired from a Chinaman named Lau Keng Tong, for five guilders a month'.[263] They sailed on 9 October 1858 touching at the islands of Tidore, March, Motir, Makian, and Kaióa before arriving on the northwest coast of the mountainous island of Batchian 13 days later.

Wallace was to remain altogether over five months on the island. He was assisted by a mining engineer named Otto Fredrik Ulrich Jacobus Huguenin (1827–1871), who had joined the Dutch East Indies government service in 1850 and by 1859 was head of the Mining Service. He lent Wallace a copy of Louis Agassiz's *Lake Superior* (1850) which Wallace noted extensively in his *Notebook 4*.[264] Huguenin's name is mentioned very briefly only twice in Wallace's notebooks and nowhere else.

There were still many unknown biological treasures to be found in this part of the world. During his stay Wallace discovered the largest bee in the

---

[263] MA2: 23–4.
[264] *Notebook 4*, p. 61b. Agassiz is noted on pages 140–7 and an unnumbered page near the back cover. Huguenin is also mentioned in *Notebook 5*, p. 3.

world—fully an inch and a half long! Wallace's Giant Bee, as it is known today, was first described by Frederick Smith in 1860: 'This species is the giant of the genus to which it belongs, and is the grandest addition which Mr. Wallace has made to our knowledge of the family Apidae'.[265] The bee was not sighted again for over 100 years. It was believed extinct until it was found again in 1981 building nests inside living termite mounds.[266]

Wallace was more excited with the capture of a large and beautiful butterfly, now called Wallace's Golden Birdwing Butterfly (*Ornithoptera croesus*). In the *Malay Archipelago* he described its capture in a celebrated passage:

[It is] one of the most gorgeously coloured butterflies in the world. Fine specimens of the male are more than seven inches across the wings, which are velvety black and fiery orange, the latter colour replacing the green of the allied species. The beauty and brilliancy of this insect are indescribable, and none but a naturalist can understand the intense excitement I experienced when I at length captured it. On taking it out of my net and opening the glorious wings, my heart began to beat violently, the blood rushed to my head, and I felt much more like fainting than I have done when in apprehension of immediate death. I had a headache the rest of the day, so great was the excitement produced by what will appear to most people a very inadequate cause.[267]

This passage is reminiscent of Wallace's 28 January 1859 letter to Stevens written shortly after the capture. The butterfly was so important that Wallace even proposed a scientific name for it.

But the greatest highlight of the collecting on Batchian was the discovery, by his assistant Ali, of a new Bird of Paradise species. Wallace mentioned it in his

---

[265] Smith, F. 1860. Catalogue of hymenopterous insects collected by Mr. A.R. Wallace in the islands of Bachian, Kaisaa, Amboyna, Gilolo, and at Dory in New Guinea. *Journal of the Proceedings of the Linnean Society: Zoology* 5 (17b): 93–143, pl.1. Smith called the bee *Megachile pluto*, currently recognized in the combination *Chalicodoma pluto*.

[266] Messer, A.C. 1984. *Chalicodoma pluto*: the world's largest bee rediscovered living communally in termite nests (Hymenoptera: Megachilidae). *Journal of the Kansas Entomological Society* 57 (1): 165–8.

[267] MA2: 51. This passage was first drafted in *Notebook 4*, p. 61.

letter of 29 October to Stevens, written just a week after his arrival on Batchian. Wallace's excitement can be gauged by the fact that he used probably the largest number of exclamation points he ever squeezed into one of his letters: 'I have got here a new *Bird of Paradise*!! of a new genus!!! quite unlike any thing yet known, very curious & very handsome!!!' It must have raised his hopes of making many other zoological discoveries. The new bird with its outstretched white feathers like banners was named by George Gray, the Ornithological Curator at the British Museum, as Wallace's standardwing, *Paradisea wallacii*.[268] No-one had suspected the existence of a Bird of Paradise so far west as Batchian.

One of Wallace's regular correspondents, keeping him informed about happenings at home, was his boyhood friend George Silk. In his letter of 30 November 1858, Wallace provides another glimpse of his attitude towards the reading of his Ternate essay at the Linnean Society meeting in July 1858. It is disarmingly casual: 'if you have any acquaintance who is a member of the Linnaean Society, borrow the Journal of *Proceedings* of August last, and in the last article you will find some of my latest *lucubrations*, with some complimentary remarks thereon by Sir C. Lyell and D$^r$ Hooker, which (as I know neither of them) I must say I am a *little* proud of'. Wallace enclosed a 350-word draft article 'Note on the smoke nuisance' written when the Society of Arts offered a gold medal for the best essay on the subject. He intended it for the weekly *Athenaeum* but it was never published. Presumably Wallace's letter arrived too late to compete for the medal.

Wallace's first letter to Darwin after hearing about the great coincidence of the convergence of their theories does not survive. We can glean something of its contents from Darwin's reply of 25 January 1859. Darwin was relieved to hear that Wallace was happy with the arrangements made on his behalf. And, perhaps of greatest interest to Wallace, he had enquired about Charles Lyell's reactions. Lyell

---

[268]  Gray, G.R. 1859. [Notes on a new bird-of-paradise discovered by Mr. Wallace.] *Proceedings of the Zoological Society of London* 27: 130. *Paradisea wallacii* is now used in the combination *Semioptera wallacii*.

was shaken but not yet convinced. Darwin referred to writing his 'abstract' of his larger work on species. This abstract is of course the *Origin of Species* (1859).

# Batchian

## 50. To Samuel Stevens, 29 October 1858

**Batchian**

*Oct' 29th. 1858*

My dear Mr Stevens[269]

As there is now a boat going which may just catch the mail at Ternate, I write a few lines to let you know of my having arrived here safe & commenced operations & to mention a few things I forgot in my last hasty letter [of 2 September 1858].

I came here in a small hired boat with my own men. Luckily it was fine weather or a hundred miles at sea with no means of cooking & only room for one day's water, would have been more than unpleasant. It was however a useful experience before my proposed more extensive voyages. I stopped 5 days at the Kaioa Islands, just half way, & got a nice coll. of beetles but few birds. A fair number of new species, & some curious varieties of those before found at Ternate & Gilolo. Here I have been also yet only 5 days,—& from the nature of the country & what I have already done, I am inclined to think it may prove one of the best localities I have yet visited.

I have already 20 species of Longicorns *new to me*, nothing very grand, but many pretty & very interesting The most remarkable is one of the Bornean genus, a *Triammatus* very near *T. cheirolatii* Pasc., also several species of the elegant little

---

[269]   The letter is annotated: 'rec'd March 9/59'.

genus *Serixia*, which have been very scarce or absent since I left Sarawak. I have also an elegant new *Pachyrhynchus*, a fine *Ips*, a small new *Cicindela* allied to the *C. funerata* Dej, of Dorey, and a small new *Therates*. In Butterflies I have taken an imperfect spec. of a glorious new species very like [Papilio] *Ulysses*, but distinct, & *even handsomer*! Perfect spec[ns] of this must be *five pounders*.

I have seen also a ♀ of a grand new Ornithoptera but cannot tell what the male will prove. *Codrus* or I rather think a new allied species I have several times seen but too wild to catch. The Macassar *Papilio* near *sarpedon* is also here, & 2-3 others I have not yet taken. Other butterflies scarce as yet but one close to or the same as the largest of the Aru *blues* very abundant. These are decidedly good prospects.

Birds are as yet very scarce but I still hope to get a fine collection though I believe I have already the finest & most wonderful bird in the island. I had a good mind to keep it a secret but I cannot resist telling you. I have got here a new *Bird of Paradise*!! of a new genus!!! quite unlike any thing yet known, very curious & very handsome!!! When I can get a couple of pairs I will send them overland to see what a new Bird of Paradise will really fetch. I expect £25 each![270] Had I seen the bird in Ternate, I would never have believed it came from here, so far out of the hitherto supposed region of the Paradiseidae. I consider it the greatest discovery I have yet made & it gives me hopes of getting other species in *Gilolo* & *Ceram*.

There is a species of monkey also here much further east than in any other island so you see this is a most curious locality combining ~~species~~ forms of the East & West of the Archipelago yet with species peculiar to itself.[271] It also differs

---

[270]   On 30 January 1859, Wallace despatched an advance consignment to Stevens which included specimens of *Ornithoptera croesus*, *Papilio telemachus*, and the new Bird of Paradise. This parcel was in transit, Ternate 6 March, Batavia 9 April, Singapore 12 April 1859 (see Baker, 2001: 278).

[271]   The monkey was described as *Cynopithecus nigrescens* by Gray, J.E. 1860. Description of a new species of Cuscus (*C. ornatus*) from the island of Batchian, with a list of the mammalia collected in that island by Mr. A.R. Wallace. *Proceedings of the Zoological Society of London* (January 11): 1–3. Wallace (MA2: 54) was correct in his assumption that the monkey was introduced to Batchian. It is now recognized as the Celebes crested macaque, *Macaca nigra* (Desmarest, 1822).

from all the other Moluccas in its geological formation containing *iron, coal, copper & Gold,* with a glorious forest vegetation & fine large mountain streams. It is a continent in miniature. The Dutch are working the coals & there is a good road to the mines which gives me easy access to the interior forests.

So much for my prospects.

1. When I mentioned subscribing to Foxcroft, I did not intend subscribing for *Lepidoptera* if separate, as I cannot afford it, & besides care little for any but the *Papilios* & *Pieridae.* Withdraw my name therefore for *Lepidoptera* after the *first* year—if you cannot at once. In *Coleoptera* I am willing to subscribe 2 years & even a third if he changes his locality, *always supposing however* that the collections are divided *fairly,* that is that each subscriber gets the *same number of species,*—the only advantage the first on the list have, being the *best* of the unique ones, & quite enough advantage too, in *fact too much,* for when *20 people* subscribe equally to anything, it is absurd that there sh^d be any advantage attached to being first on the list.

2. I see among the pins last sent some *no. 14* which are perfectly useless being so short. They are as short as 7, & as thick as 13. What I want is a pin nearly as fine as 7, but *much longer* because many delicate beetles have such long legs that 7 does not raise them high enough. This sort of pin I hope to find in the next box as I see you have sent some no. 5 (I think).

3. An Essay on *varieties* which I sent to Mr. Darwin has been ~~presented~~ read to the Linnaean Soc. by D^r Hooker & Sir C. Lyell on account of an extraordinary coincidence with some views of Mr. Darwin, long written but not yet published, & which were also read at the same meeting. If these are published I dare say Mr. Kippist will let you have a dozen copies for me.[272] If so send me 3, & of the remainder send one to *Bates, Spruce,* & any other of my friends who may be interested in the matter.

---

[272]  The botanist Richard Kippist (1812–1882) was librarian of the Linnean Society of London (1842–1881). This was published as Wallace's Ternate essay, 1858.

4. Again I wish to remind you to obtain *2 copies* of all the papers published under my agreement with M^r Saunders sending me *one* & keeping the other carefully for me. I should much like also a copy of Mr. Westwoods paper on the Singapore *Cleridae* as many of the same genera & closely allied species occur all over the Archipelago.[273]

I do not think *Siam* much of a locality. It is too much cultivated. *Cambodia* on the other hand is a grand country for a naturalist. I know much about it from the French missionaries I met at Singapore.

Everybody says *Ceram* is a fine country for birds but I believe *no* remarkably fine birds are known from it but *Lories &c.* The natives all say the birds of Ceram are very beautiful & they judge *only* by the domesticated birds, *lories, cockatoes* &c. It may be very good but I think there is no good grounds for saying *it is so.*

What a wonderful locality Ega is! for the museum to take £70 worth from a collection after 2 years previous collecting there. Two months exhausts my localities here.

I enclose a note for my mother which please forward. I can do nothing at drawing birds, but send a horrible sketch of my *discovery* that you may not die of curiosity.[274] I am told the wet season here is terrible & it begins in Dec^r, so I shall probably have to leave but if I do well, may return next year two months

[273] Westwood, John Obadiah. 1855. Descriptions of some new species of Cleridae, collected at Singapore by Mr. Wallace. *Proceedings of the Zoological Society of London* (13 February): 19–26, pl. XXXVIII.

[274] Wallace's sketch of the bird has not been found. It was examined by George R. Gray at the British Museum, who proposed a new name for the bird: 'I have endeavoured to transform the rough sketch into the probable appearance of the living bird; and I further add the provisional specific name of *Paradisea wallacii*, which appellation I think is justly due to Mr. Wallace for the indefatigable energy he has hitherto shown in the advancement of ornithological and entomological knowledge, by visiting localities rarely if ever travelled by naturalists'. Gray, G.R. 1859. [Notes on a new bird-of-paradise discovered by Mr. Wallace.] *Proceedings of the Zoological Society of London* (22 March): 130.

earlier. I am waiting anxiously to hear that the Amboyna beetles have come safe & unspoilt.

I must now remain

Yours very faithfully

Alfred R. Wallace

*Samuel Stevens* Esq.

---

## 51. To Francis Polkingthorne Pascoe, 28 November 1858

Batchian

*Nov. 28th. 1858*

*F.P. Pascoe*, Esq.

My dear Sir

It was with great pleasure I heard from Mr Stevens some time since that you proposed publishing a complete list of my Aru longicorns, & it was with much regret I learnt from a subsequent letter that you had abandoned your intention in consequence of Mr. Thompson having figured a few of the finest sp. in his Archives.[275] I really think you sh$^d$ have got the start of him as you had or might have had my private coll. long before he got his sets. However that may be I trust that on the arrival of my *Dorey* insects you will again take up the matter & by combining the Aru & Dorey coll$^{ns}$ with those of the French & Dutch naturalists give us a complete Synopsis of the Longicorns of the New Guinea fauna yet known, a labour that will /resound/ to your credit & be far more

---

[275] Thomson, J. 1857. Wallace, voyage dans l'Asie Orientale: fragments entomologiques renfermant la description de coléoptères nouveaux ou rares. *Archives Entomologiques* 1: 425–60. See Hayek, C.M.F. von. 1989. A short biography of the entomologist James Thomson and the dates of publication of the Archives Entomologiques, Arcana Naturae, Monographie des Cicindélides, Musée Scientifique and Physis. *Archives of Natural History* 16 (1): 81–99.

interesting & useful to entomologists than any quantity of isolated descriptions & figures of even the most magnificent species.

Allow me also to observe (& pray do not take offence at the observation) that you have placed me in the same (or a worse) position as Mr. T. [Thomson] has placed you,—by describing only the finer & more interesting new species from my Malacca & Borneo Longicorns leaving me a mass of minute & obscure species which I shall have the labour of working out & describing myself. This you may imagine is not an agreeable prospect & I therefore told Mr. Stevens to inform you of a rule I had long since made, not to lend any portions of my private coll$^{ns}$ except to parties who would work out entire groups from one or more localities.

As I look forward to publishing on my return a complete synopsis of the Col$^a$ of the Archipelago, it is necessary for me to obtain as much previous assistance as possible,—besides wh. I object altogether *on principle* to picking out the more interesting & well marked new sp. from a collection, for publication. Every one however has a right to his own opinions on this matter & can follow his own course as regards the sp. in his own collection. All I require is that my finer *unique* sp. should either be taken with the entire group to which they belong or left to season my own labour on the unsavoury mass of minute obscurities.

In the Dorey coll. are a good many fair & interesting forms mostly unique or rare. The Tmesisterni are very difficult & I think it is

Fig. 19. Francis Polkinghorne Pascoe

impossible to determine them without comparison with the N. Guinea speci-mens in the Paris & Leyden coll[ns]. *T. septempunctatus* Bois. I found at Dorey. His fig. in "Voy. of Astrolabe" is wretched & his locality, *Amboyna* (as usual) wrong.[276]

My small Amboyna coll. contains some lovely things & at Gilolo & this place I am getting lots of new & pretty species.

I remain,

my dear Mr Pascoe,

Your Sincerely,

Alfred R. Wallace

F.P. Pascoe Esq.

## 52. To George Silk, 30 November 1858

### Batchian, Moluccas

*Nov. 30th. /58*

My dear George

I do not think I have written to you very lately. I have just received yours of August 3 with reminiscences of Switzerland.[277] To you it seems a short time since. To me an immeasurable series of ages. In fact, Switzerland & the Amazon now seem to me quite unreal,—a sort of former existence,—a long-ago dream. Malays & Papuans,—Beetles & Birds, are what now occupy my thoughts, mixed with financial speculations & hopes for a happy future in Old England, where I may live in solitude & seclusion, except from a few choice friends.

You cannot, perhaps, imagine how I have come to love solitude. I seldom have a visitor but what I wish him away in an hour. I find it very favourable to

---

[276]  Boisduval, J.B.A.D. ed. 1832–1835. *Voyage de l'Astrolabe. Faune entomologique de l'Océanie*, 2 vols.

[277]  Wallace and Silk travelled to Switzerland in Autumn 1853. ML1: 325.

reflection, & if you have any acquaintance who is a member of the Linnaean Society, borrow the Journal of *Proceedings* of August last, & in the last article you will find some of my latest *lucubrations*, with some complimentary remarks thereon by Sir. C. Lyell & D^r Hooker, which (as I know neither of them) I must say I am a *little* proud of.[278]

As to politics, I hate & abominate them. The news from India I now never read, as it is all an inextricable confusion without good maps & regular papers, mine come in lumps 2 or 3 months at a time, with the alternate ones stolen or lost. I therefore beg you to write no more politics. Nothing public or newspaperish. Tell me about yourself,—your own private doings,—your health, your visits, your new or old acquaintances (for I know you pick up ½ a dozen every month, *à la Barragan*).

But, above all, tell me of what you read. Have you read the Currency book I returned you?[279]—Horne Tooke?—Bentham? Family Herald Leading Articles?—Give me your opinions on any or all of these. Follow the advice in Fam. Herald art. on 'Happiness'—'*Ride a hobby*', & you will assuredly find happiness in it, as I do. Let *Ethnology* be your hobby, as you seem already to have put your foot in the stirrup,—but *ride it hard*. If I live to return I shall come out strong on Malay & Papuan races, & shall astonish Latham, Davis, &c &c.

By-the-by, I have just had a letter from Davis; he ~~says he~~ says he sent my last letter to you, & it is lost mysteriously. Instead, therefore, of sending me an answer to my *poser*, he repeats what he has said in *every letter* I have had from

---

[278]   Wallace, Ternate essay, 1858.

[279]   In this paragraph, Wallace recommends a range of literature to Silk. The 'Currency' book might have been Combe, G. 1852. *The currency question*. London. The others are: Horne Took, J. *Epea pteroenta, or, the diversions of Purley*; works by Jeremy Bentham (1748–1832); articles in *The Family Herald*; Latham, R.G. 1854. *The natural history department of the Crystal Palace described: ethnology*; works by Joseph Barnard Davis (1801–1881); Prichard, J.C. *The natural history of man*; Lawrence, W. *Lectures on physiology, zoology, and the natural history of man*; Sterne, L. *Tristram Shandy*; Dumas, A. *Queen's Necklace*; Dumas, A. *The Memoirs of a physician*.

him, "*myriads of miracles are required to people the earth from one source.*" I am sick of him. You must read "Pritchard" through, & "Lawrence's Lectures on Man" carefully, but I am convinced no man can be a good ethnologist who does not travel, & not travel merely, but reside, *as I do*, months and years with each race, becoming well acquainted with their average physiognomy & moral character, so as to be able to detect *cross-breeds*, which totally mislead the hasty traveler, *who thinks they are transitions*!! Latham, I am sure, is quite wrong on many points.

To New Guinea, I took an old edition of "*Tristram Shandy*" which I read through about *three times*. It is an annoying & you will perhaps say, a very gross book, but there are passages in it that have never been surpassed, while the character of Uncle Toby has, I think, never been equaled, except perhaps by that of Don Quixote. I have lately read a good many of Dumas's wonderful novels, & they *are wonderful,* but often very careless & some quite unfinished. The "Memoirs of a Physician" is a most wonderful wild mixture of history, science, & romance—the 2nd part, "*The Queen's Necklace*" is a most wonderful & perhaps most true. You should read them (if you have not) when you are horribly "ennui".

As to your private communications in former letters, I am very sorry you have not been fortunate in your "affaires du coeur". All I can say is "try again". Marriage has a wonderful effect in brightening the intellect. For example, John [Wallace] used not to be considered *witty*, yet in his last letter he begs me "to write to him *semi-occasionally, or oftener if I have time*, & I send you a not bad extract from his letter, with an idea of my own on "smoke" *to send to the Athenaeum.* By this mail I send more than a dozen letters, for my correspondence is increasing. You must therefore excuse this random list of odds & ends & send me a ditto in return only one day.

I must now conclude

Remaining my dear G.,

Yours ever faithfully,

Alfred R. Wallace

*G. C. Silk* Esq.

P.S. A big spider fell close to my hand in the middle of my signature wh. accounts for the hitch.

P.S. I have to send this at a moments notice. Can not write home, so call on my mother. AW. Dec.20

[Enclosure 1]

*Note on the smoke nuisance*

How is it that amid the lamentations & grumbling over the incalculable mischief done by London smoke,—masterpieces of art ruined, palaces spoilt before they are finished, life & property lost in November fogs, our streets & squares & noblest public buildings all rendered hideous, our clothes & persons begrimed & our lungs diseased.—

There should be no proposals made to go to the fountain head & instead of recovering our galleries & museums to a distance from those who most want them, *try to get rid of the smoke itself.* When the thing is once done, where our city is clean, our skill bright our air pure our linen unsoiled & our works of art uninjured, we shall be almost incredulous that such a state of apathy and barbarism could ever have existed.

The thing can easily be done;—it is a mere matter of cost, & the expense of rendering each house in London smokeless it is not very difficult to calculate. We have the choice of gas, anthracite coal, or of substituting Arnotts or any other smokeless grates & cooking ranges for those now in use, either of which if not absolutely perfect would certainly get rid of nine tenth of the smoke now produced, & would probably soon repay the expense of the changes in the saving of fuel. What hardship, what impossibility,

The factories have been parliamentarized but the million domestic hearths are at once more mischievous & easier to deal with. What interference with vested right would there be in compelling by Act of Parliament the use of either of these methods, any more than in compelling chimneys to be swept at certain intervals or houses to be built of a certain stability? Why, the mere saving in soap & linen would cover the expense in a few years, to say nothing of the incalculable national & sanitary advantages already alluded to.

If the Athenaeum & the Times would vigorously take up the question we might yet see our whole city not only the largest & the wealthiest but the cleanest & the healthiest in the world.

Alfred R. Wallace

Batchian Moluccas

*Nov. 1858*

[Enclosure 2]

*How to manage matters in the Model Republic*

Extract from a letter from California [by John Wallace]. "I must tell you that a friend of mine Mr. Mandeville has been appointed by the President to the office of the U.S. Surveyor General for the State of California.[280] You will probably imagine that in a Model Republic like that of the United States an office of this kind would be filled by a person well qualified in every respect for such an important post; but the fact is Mr. Mandeville knows nothing either theoretically or practically of surveying,—has not even the remotest idea of the first principles; but there *he has been an active politician & is on the winning side*. This is the way all appointments to offices are made in this Country, no questions as to qualification asked, & no "right-man-in-the-right-place" cry raised. The principle seems to be "to the victors belong the spoils," and an active man of the right party is considered to be well qualified for *any office!*"

---

[280]   James W. Mandeville (1824–1876) was a politician, elected as US Surveyor General for the District of California for 1858–1859. The President of the United States of America for the term of 1857–1861 was James Buchanan (1791–1868).

via *Southampton*.[281]

G.C. Silk Esq.

79, Pall Mall

*London W.C.*

―――

## 53. From Charles Darwin, 25 January 1859

**Down Bromley Kent**

*Jan. 25ᵗʰ*

My dear Sir

I was extremely much pleased at receiving three days ago your letter to me
& that to Dʳ Hooker. Permit me to say how heartily I admire the spirit in
which they are written. Though I had absolutely nothing whatever to do in
leading Lyell & Hooker to what they thought a fair course of action, yet I
naturally could not but feel anxious to hear what your impression would be.
I owe indirectly much to you & them; for I almost think that Lyell would
have proved right & I shᵈ never have completed my larger work, for I have
found my abstract hard enough with my poor health, but now thank God I
am in my last chapter, but one. My abstract will make a small vol. of 400 or
500 pages.—Whenever published, I will of course send you a copy, & then
you will see what I mean about the part which I believe Selection has played

---

[281]  There are three postmarks around the address. The first reads: FRANCO. ZEE
BRIEF. TERNATE. The second (subsequently cancelled in ink, probably by
the postmaster as it contradicts the first postmark): ONGEFRANKEERD.
ZEE BRIEF. TERNATE. The third: LONDON BV AP 6 59. Postmarks in
Dutch. 'Franco' is prepaid. 'Zee Brief' is 'letter by sea'. 'Ongefrankeerd' is
unstamped, where the recipient pays the postage.

with domestic productions. It is a very different part, as you suppose, from that played by "Natural Selection".—

I sent off, by same address as this note, a copy of Journal of Linn. Soc. & subsequently I have sent some ½ dozen copies of the Paper.[282]—I have many other copies at your disposal; & I sent two to your friend D[r] Davies(?) author of works on men's skulls.[283]—

I am glad to hear that you have been attending to Bird's nest; I have done so, though almost exclusively under one point of view, viz. to show that instincts vary, so that selection could work on & improve them. Few other instincts, so to speak, can be preserved in a museum.—

Many thanks for your offer to look after Horses stripes; if there are any Donkey's pray add them.—

I am delighted to hear that you have collected Bees' combs; when next in London I will enquire of F. Smith & Mr Saunders.[284] This is an especial hobby of mine, & I think I can throw light on subject.—If you can collect duplicates at no very great expense, I sh[d] be glad of specimens for myself with some Bees of each kind.—Young growing & irregular combs, & those which have not had pupæ are most valuable for measurements & examination: their edges sh[d] be well protected against abrasion.—

Everyone whom I have seen has thought your paper very well written & interesting. It puts my extracts, (written in 1839 now just 20 years ago!) which I must say in apology were never for an instant intended for publication, in the shade.

You ask about Lyell's frame of mind. I think he is somewhat staggered, but does not give in, & speaks with horror often to me, of what a thing it would be

---

[282] Wallace, Ternate essay, 1858.

[283] Joseph Barnard Davis (1801–1881) was the author of *Crania Britannica* (1856–1865).

[284] William Wilson Saunders (1809–1879) and Frederick Smith (1805–1879), assistant keeper in the Zoology Department of the British Museum, specializing in Hymenoptera.

& what a job it would be for the next Edition of Principles, if he were "perverted".—But he is most candid & honest & I think will end by being perverted. D$^r$ Hooker has become almost as heterodox as you or I.—and I look at Hooker as by far the most capable judge in Europe.—

Most cordially do I wish you health & entire success in all your pursuits & God knows if admirable zeal & energy deserve success, most amply do you deserve it.

I look at my own career as nearly run out: if I can publish my abstract & perhaps my greater work on same subject, I shall look at my course as done.

Believe me, my dear Sir

Yours very sincerely

C. Darwin

## 54. To Samuel Stevens, 28 January 1859

### Batchian, Moluccas

*January 28, 1859*

I had determined to leave here about this time, but two circumstances decided me to prolong my stay—first, I succeeded at last in taking the magnificent new Ornithoptera, and, secondly, I obtained positive information of the existence here of a second species of Paradisea, apparently more beautiful and curious than the one I have obtained. You may perhaps imagine my excitement when, after seeing it only two or three times in three months, I at length took a male Ornithoptera. When I took it out of my net, and opened its gorgeous wings, I was nearer fainting with delight and excitement than I have ever been in my life; my heart beat violently, and the blood rushed to my head, leaving a head-ache for the rest of the day.

The insect surpassed my expectations, being, though allied to Priamus, perfectly new, distinct, and of a most gorgeous and unique colour; it is a fiery golden orange, changing, when viewed obliquely, to opaline-yellow and green. It is, I think, the finest of the Ornithoptera, and consequently the finest butterfly in the world? Besides the colour, it differs much in markings from all of the Priamus group. Soon after I first took it I set one of my men to search for it daily, giving him a premium on every specimen, good or bad, he takes; he consequently works hard from early morn to dewy eve, and occasionally brings home one; unfortunately several of them are in bad condition. I also occasionally take the lovely Papilio Telemachus, n.s.

I have sent off a small box containing four males, one female, and one young bird of the new Batchian Paradisea, besides one red-ticketed private specimen; six males and five females of the new Ornithoptera, and seven Papilio Telemachus.

Tell Mr. Gray and Mr. Gould that the Paradisea had better not be described yet, as I am making great exertions to get the second species, evidently of the same genus, which will enable a generic character to be more accurately given. The butterflies, I trust, will be both figured, male and female, either in Mr. Hewitson's[285] book or in Ent. Soc. Trans. For the Ornithoptera I propose Crœsus as a good name.

Butterflies are scarce; good beetles turn up occasionally, but nothing very grand. I have now a handsome series of Buprestidæ, and a remarkably pretty lot of Longicorns; one of my last acquisitions is a grand bronzy Tmesisternus, 1 ½ inch long, a single specimen only. In almost all orders, and in birds, there is a deficiency of species; yet there are so many pretty and brilliant things, and a few so grand and new, that on the whole I am inclined to think my Batchian collection will be the best I have made anywhere.

Another reason which may induce me to stay perhaps two or three months longer at Batchian is that I have had no fever here, which I have never been free from two months at a time for the last two years before; and I may

---

[285] Hewitson, W.C. 1846–1852. *The genera of diurnal lepidoptera*, 3 vols. London.

therefore hope to get my health well established for my next journey to New Guinea.

The butterflies will make a show-box which will, I think, be admired almost as much as the birds of Paradise.

---

## 55. From Charles Darwin, 6 April 1859

**Down Bromley Kent**

*April 6 /59*

My dear Mr Wallace

I this morning received your pleasant & friendly note of Nov. 30th. The first part of my M.S is in Murray's hands to see if he likes to publish it. There is no preface, but a short Introduction, which must be read by everyone, who reads my Book. The second Paragraph in the Introduction, I have had copied *verbatim* from my foul copy, & you will, I hope, think that I have fairly noticed your paper in Linn. Transacts—You must remember that I am now publishing only an Abstract & I give no references. I shall of course allude to your paper on Distribution; & I have added that I know from correspondence that your explanation of your law is the same as that which I offer.[286]—You are right, that I came to conclusion that Selection was the principle of change from study of domesticated productions; & then reading Malthus I saw at once how to apply this principle.[287]—Geographical Distrib. & Geological relations of extinct to

---

[286]  In *On the origin of species* (1859, p. 355), Darwin noted 'I now know from correspondence, that this coincidence [Wallace] attributes to generation with modification'.

[287]  Malthus, T. 1826. *An essay on the principle of population.* London. This is the first mention of Malthus in the extant Darwin–Wallace correspondence.

recent inhabitants of S. America first led me to subject. Especially case of Gala-pagos Isl<sup>ds</sup>.—

I hope to go to press in early part of next month.—It will be small volume of about 500 pages or so. I will of course send you a copy.

I forget whether I told you that Hooker, who is our best British Botanist & perhaps best in World, is a *full* convert, & is now going immediately to publish his confession of Faith; & I expect daily to see the proof-sheets.—Huxley[288] is changed & believes in mutation of species: whether a *convert* to us, I do not quite know.—We shall live to see all the *younger* men converts. My neighbour & excellent naturalist J. Lubbock[289] is enthusiastic convert.

I see by Nat. Hist. notices that you are doing great work in the Archipelago; & most heartily do I sympathise with you. For God sake take care of your health. There have been few such noble labourers in the cause of Natural Science as you are.

Farewell, with every good wish

Yours sincerely

C. Darwin

P.S. You cannot tell how I admire your spirit, in the manner in which you have taken all that was done about publishing our papers. I had actually written a letter to you, stating that I would not publish anything before you had published. I had not sent that letter to the Post, when I received one from Lyell & Hooker, urging me to send some M.S. to them, & allow them to act as they thought fair & honourably to both of us. & I did so.—

---

[288]  Thomas Henry Huxley (1825–1895), who after a navy career became professor of Natural History at the Royal School of Mines.

[289]  John Lubbock (1834–1913), a banker with varied interests in natural history. He grew up at High Elms Estate, close to Down House.

# Ternate

## 56. To Thomas Sims, 25 April 1859

**Ternate**

*April 25th. 1859*

My dear Thomas

Many thanks for your long letter which is the best you have ever written me, but I am too busy to promise you much in return.

Your ingenious arguments to persuade me to come home are quite unconvincing. I have much to do yet before I can return with satisfaction of mind;—were I to leave now I should be ever regretful & unhappy. That alone is an all-sufficient reason. If you like, I feel my work is here as well as my pleasure & why should I not follow out my vocation? As to materials for work at home, you are in error. I have indeed materials for a life's study of Entomology, as far as the forms & structure & affinities of insects are concerned,—but I am engaged in a wider & more general study—that of the relations of animals to time & space, or in other words their Geographical & Geological distribution & its causes.

I have set myself to work out this problem in the Indo-Australian Archipelago & I must visit & explore the largest number of islands possible & collect materials from the greatest number of localities in order to arrive at any definite results. As to health & life, what are they compared with peace & happiness, & *happiness* is admirably described in the *Fam. Herald* as obtained by "work with a purpose, & the nobler the purpose the greater the happiness."—But besides these weighty reasons there are others quite as powerful,—pecuniary ones. I have not yet made enough to live upon, & I am likely to make it quicker here than I could in England. In England there is only one way in which I could live, by returning to my old profession of surveying. Now, though I always liked surveying, I like collecting better, & I could never now give my whole mind to any work apart from the study to which I have devoted my life.

So far from being angry at being called an Enthusiast, it is my pride & glory to be worthy to be so called. Who ever did anything good or great who was not an enthusiast? The majority of mankind are enthusiasts only in one thing, in money-getting; & these call others enthusiasts as a term of reproach, because they think there is something in the world better than money-getting. As to riding in carriage &c. it strikes me that the power or capability of getting rich is in an *inverse proportion* to a man's reflective powers & in *direct proportion* to his impudence. It is good to be rich, but not to get rich, or to try to get rich, & few men are so unfitted to get rich, if they did try as myself.

I do not understand your attack on Claudet.[290] I did not quite see what his doctrine was, but I saw an ingenious combination to produce the stereoscopic effect by looking at a single ~~picture~~ image,—& if that effect is produced, he is entitled to all credit for the discovery. You shirk this question altogether. If three or four persons can stand before a ground glass & on it can all see at once a picture as solid as in the stereoscope, here is an ingenious practical discovery.—

I saw a notice also of another means of producing the same effect by a most ingenious combination. Two stereoscopic pictures are thrown by magic lanthern on one disc so as exactly to overlap, but one is ~~passed~~ covered with red the other with green glass (or any two complementary colours) so as to form two coloured images which to the naked eye will be white & slightly confused, but the *spectators* all *wear spectacles*, one eye *green* the *other red*, & the consequence is that one eye sees only one picture, & the stereoscopic effect is produced! The idea of this highly delights me & I hope it is successful.

Your idea of painters painting with two eyes & producing stereoscopic effects I do not understand, but if there is anything in it I suggest this experiment. Have a camera with two object glasses, the centre, the distance apart of the two eyes & adjusted to throw their focal image exactly on the same surface. Thus,—[fig. 20] take pictures with this camera & you should have the effect of nature seen with two eyes.—If the result was only to blur the outlines & near the effect, the

---

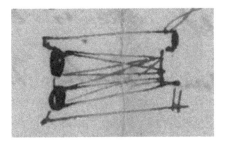

Fig. 20. Wallace's figure illustrating his letter to Sims of 25 April 1859

painters must paint with one eye only, & it is evident that for every thing but near objects our two eyes have but the effect of one,—& this is the reason why landscape stereo-scopes have none of the charm of small close objects which produce such a startling effect of solid reality. I think for a *distant* landscape two *identical* pictures in the stereoscope would produce an equally good effect. Have you ever tried it? For it evident that practically the two eyes see distant objects absolutely the same while close objects they see very different. I hope your engraving plans may succeed & be profitable. I know not exactly where I am going next, but perhaps shall before I close this. In haste,

Yours very affectionately,

Alfred R. Wallace.

Thos. Sims, Esq.

## 57. From Daniel Hanbury, 6 May 1859

**London**

*6 May 1859*

My dear Sir—

I have rec[d] your note dated Ternate Oct. 5, 1858, and as Mr. Stevens offers to enclose a few lines by way of reply I avail myself of the opportunity.

Your remarks about the Laurineous Barks I named, being produced in the interior of New Guinea appears to preclude all chance of determining their botanical origin at present. But as it is something to know with certainty even the locality where these drugs are produced, I sh[d] be glad to have samples if you have convenient opportunity for procuring them and if they will pack without detriment to other things.

Your ornithological discoveries are extremely interesting, even to those who ~~not professed~~ have no claim to be called ornithologists. For myself I may say that even from a child, I have regarded the B^s. of Parad. with especial interest and years and years ago delighted in attempts to delineate their varied forms.

Yours &c.

D.H.

A.R. Wallace Esq.

Fig. 21. Daniel Hanbury

# MENADO, AMBOYNA, AND CERAM

## 10 June 1859–16 June 1860

Wallace left Batchian on 13 April 1859 on a large Dutch government double outrigger called a coracora. During the journey he narrowly escaped being bitten by a snake in his cabin. Wallace arrived back on Ternate on 19 April to sort and pack his collections. He stayed just under a fortnight and then decided to undertake a journey around the Moluccan islands, taking a steamer (on 1 May 1859) south to Ambon, Banda, and Timor and then back north to Menado in the northeastern part of Celebes. He landed at the Dutch town of Menado on 10 June. Apparently his fathful servant Ali stayed behind in Ternate, where he had just married, but continued collecting for Wallace on Gilolo.

Around this time Wallace wrote an important article: 'On the zoological geography of the Malay Archipelago'. He sent it to Darwin to forward to the Linnean Society where it was read on 3 November 1859. In it Wallace elaborated on his ideas about the division of the Archipelago into sharply delineated Asiatic and Australian regions.

Wallace would stay over three months in the country around Menado. He was particularly interested in observing the island's endemic Maleo, or Brush-turkey (*Macrocephalon maleo*). He saw how the birds buried their large eggs in the sand on isolated beaches and had feet that were 'slightly webbed at the base, and thus the whole foot and rather long leg are well adapted to scratch away rapidly a loose sand'.[291] When he described the birds in a paper written in October 1859, he mentioned one of his favourite theoretical concerns—whether behaviour precedes and thus determines the structure of animals or vice versa:

---

[291]  Wallace, A.R. 1860. The ornithology of Northern Celebes. *Ibis* 2 (6): 140–7 [S57].

'For a perfect solution of the problem we must, however, have recourse to Mr. Darwin's principle of "natural selection," and need not then despair of arriving at a complete and true "theory of instinct."' This is the first time since the Ternate essay that Wallace mentioned natural selection in print. It is also the first time he used this phrase and also the first time that he attributed it solely and wholeheartedly to Darwin, omitting his own input—even before the *Origin of species* was published.

Wallace's collections in Menado completed, he boarded a steamer on 23 September heading, unusually, in a clockwise direction towards Ternate. He did not linger, but continued on the steamer to Amboyna, where he arrived on 29 September. He rented a small house near the beach for a month to pack his collections, write letters, and prepare for his next expedition to the larger island of Ceram to the north.

One of his correspondents had written to him from London that both John Gould and George Gray, arguably the most eminent ornithologists in the Western world, had recommended a visit to Ceram because the birds there were 'very fine'. How could Wallace resist such an appealing introduction? He was to stay over nine months on Amboyna and Ceram, although partly housebound due to sickness, from the end of September 1859 to June 1860. Although he tried several collecting grounds in the southwestern and eastern parts of Ceram, he was continually disappointed: 'all was barren; birds were scarcer than ever; and the natives were quite astonished at being asked about handsome birds, assuring me they knew of none in their country'.[292] With all hopes of grand zoological discoveries dashed, he must have felt poorly rewarded for his time and efforts.

After his return to Amboyna from Ceram on 31 December 1859 Wallace received from Darwin a copy of the *Origin of species*, published in November. It was sent care of Hamilton, Gray and Co. in Singapore.[293] Wallace must have

---

[292]  Wallace, A.R. 1861. On the ornithology of Ceram and Waigiou. *Ibis* 3 (11): 283–91, pl. IX [S62].

[293]  See CCD8: 556. Wallace's annotated copy of *Origin* survives at Cambridge University Library. See Beddall, B.G. 1988.

been deeply fascinated to at last see Darwin's views and evidence in detail.
Wallace would read the book through again and again. At the back of an offprint
of their Linnean article he jotted some initial reactions to reading Darwin's book:

*1860. Feb.*

After reading Mr Darwin's admirable work *"On the Origin of Species"*, I find that there
is absolutely nothing here [in the Ternate essay] that is not in almost perfect agreement
with that gentlemans facts & opinions.

His work however touches upon & explains in detail many points which I had scarcely
thought upon,—as the *laws of variation, correlation of growth, sexual selection, the origin
of instincts & of neuter insects,* & *the true explanation of Embryological affinities.* Many of
his facts & explanations in Geographical distribution are also quite new to me & of the
highest interest—

*ARWallace..Amboina.*[294]

Wallace was obviously excited about Darwin's work and would continue to
praise it in similar terms for the remainder of his life.

# Menado

## 58. To Francis Polkingthorne Pascoe, 20 July 1859
### Menado, N. Celebes
*July 20th. 1859*

My dear Sir,

In reply to your note of Apr. 18. allow me to assure you, that I never imagined
you knew of my wishes as to my private collection, & only meant to inform you

---

[294] NHM WP7/9 21. A transcription of this item was first published by Beccaloni
in: Smith and Beccaloni. 2008. *Natural selection and beyond,* pp. 96–7.

that I was making no special exception in your case. The express stipulations I had made with Mr Saunders to have *complete lists* published of all those parts of my private coll$^n$ I gave up to him, & also of any parts of the Coleoptera he could make arrangements for having worked out, would I imagined have made my intention apparent to Mr. Stevens without a special intimation as to lending my private coll$^n$ which however I imagined (I suppose erroneously) I had also given him.

I am not aware that to this day any one has published selected species from my private coll$^n$ but yourself. Mr. Tatum[295] never did any thing but put mss. names to a few sp. of Geodephaga & gave me some information wh. I requested from him. Mr. Baly[296] has I know long been at work on a monograph of the *Hispidae* [beetles] for which I presume he has had the specimens you refer to. In groups of little difficulty as the *Cicindelidae* [tiger beetles] & *Papilios* [butterflies] I sh$^d$ have no objection to new sp. alone being described & figured, but the Longicorns are a most difficult & complicated group, & I want assistance in working them out.

I am still in hopes you may take up some of the genera or give a complete catalogue of some local collection. An excellent opportunity of this kind occurs in my Batchian coll$^n$ wh. will probably soon after you receive this. It is I believe perfectly virgin ground, as I am not aware of any insect being known from the island before & a list of any family would therefore possess a general interest. I obtained about 150 species,—many very curious forms & some very fine insects, among them I think the handsomest *Glenea* known & a grand new *Tmesisternus*. Of a considerable number the two sexes are determined which adds much to the completeness of the coll$^n$.

---

[295]  Thomas Tatum (1802–1879), surgeon at St George's Hospital, was a member of the Royal Entomological Society of London.

[296]  Joseph Sugar Baly (1817–1890) published a *Catalogue of Hispidae in the collection of the British Museum* in 1858.

It will give me very great pleasure to hear that you will undertake to make a complete list of these, with descriptions of the new species.[297]

There occurs an extraordinary case of two *Gleneas twice* taken "in copula"— yet I cannot consider them as one species because they are the most dissimilar of the series. One of the *pairs* taken you will see in my private collection, & I invite your opinion on the matter. I may remark that in *every case* in which specimens are marked ♂. ♀. they have been taken "in cop." Where paired from several characters I merely place the specimens together without marking them as ascertained sexes.

You are no doubt aware that my hobby is "Geog. distribution",—which I am glad to hear, you are also interested in,—but it is for that very reason that I am so anxious that my collection should be worked at, either in classificational or geographic *groups*. Is there any hopes of one getting the rest of White's B.M. Cat. of the Longicorns?[298] Catalogues published in that fragmentary manner are almost useless as the first parts become quite obsolete by the time the last are published. I hope Lacordaire's vol. on the family will be out soon.

A line from you when convenient will give me much pleasure.

I must now remain

Yours faithfully

Alfred R. Wallace.

*F. Pascoe* Esq.

---

[297] Between 1864 and 1869 Wallace helped Pascoe to complete the Longicornia Malayana published in *Transactions of the Entomological Society of London* (series 3) 3: 1–712.

[298] Adam White (1817–1878) worked at the British Museum specializing in insects and crustaceans. He wrote several parts of the *Catalogue of coleopterous insects in the collection of the British Museum*. London, 1847–1855.

## 59. From Charles Darwin, 9 August 1859

**Down Bromley Kent**
*Aug.̇ 9, 1859*

My dear Mr Wallace

I received your letter & memoir on the 7$^{th}$ & will forward it tomorrow ~~or day after~~ to Linn. Soc$^y$.[299] But you will be aware that there is no meeting till beginning of November. Your paper seems to me *admirable* in matter, style & reasoning; & I thank you for allowing me to read it. Had I read it some months ago I sh$^d$. have profited by it for my forthcoming volume.—But my two chapters on this subject are in type; & though not yet corrected, I am so wearied out & weak in health, that I am fully resolved not to add one word & merely improve style. So you will see that my views are nearly the same with yours, & you may rely on it that not one word shall be altered owing to my having read your ideas.

Are you aware that Mr W. Earl published several years ago the view of distribution of animals in Malay Archipelago in relation to the depth of the sea between the islands?[300] I was much struck with this & have been in habit of noting all facts on distribution in the Archipelago & elsewhere in this relation. I have been led to conclude that there has been a good deal of naturalisation in the different Malay islands & which I have thought to certain extent would account for anomalies.

---

[299] Wallace, A.R. 1860. On the zoological geography of the Malay Archipelago. *Journal of the Proceedings of the Linnean Society: Zoology* 4: 172–84.

[300] George Windsor Earl (1813–1865) published (1845): On the physical structure and arrangement of the islands in the Indian Archipelago. *Journal of the Royal Geographical Society* 15: 358–65. Earl divided the Archipelago into eastern and western zones based on geology and ocean depths, showing that the islands were connected by shallow shelves. Wallace apparently did not read Earl until Darwin pointed it out. Wallace first cited Earl in Wallace, A.R. 1863. On the physical geography of the Malay Archipelago. *Journal of the Royal Geographical Society* 33: 217–34 [S78].

Timor has been my greatest puzzle. What do you say to the peculiar Felis there?[301] I wish that you had visited Timor: it has been asserted that fossil Mastodon or Elephant's tooth (I forget which) has been found there, which would be grand fact.—I was aware that Celebes was very peculiar; but the relation to Africa is quite new to me & marvelous & almost passes belief.[302]—It is as anomalous as relation of plants in S. W. Australia to Cape of Good Hope.

I differ <u>wholly</u> from you on colonisation of ocean's islands, but you will have *everyone* else on your side. I quite agree with respect to all islands not situated far in ocean. I quite agree on little occasional intermigration between lands when once pretty well stocked with inhabitants, but think this does not apply to rising & ill-stocked islands.

Are you aware that *annually* birds are blown to Madeira, to Azores, (& to Bermuda from America).—[303] I wish I had given fuller abstract of my reasons for not believing in Forbes' great continental extensions; but it is too late, for I will alter nothing. I am worn out & must have rest.—

[Richard] Owen, I do not doubt, will bitterly oppose us; but I regard this very little; as he is a poor reasoner & deeply considers the good opinion of the world, especially the aristocratic world.—

Hooker is publishing a grand Introduction to Flora of Australia & goes the whole length.[304]—I have seen proofs of about half.—

---

[301]  *Felis megalotis,* a striped cat from Timor, is a kind of feral house cat mentioned by Wallace (MA1: 326).

[302]  Wallace suggested that Celebes had once been connected with Africa, in: Wallace, A.R. 1859. Letter from Mr. Wallace concerning the geographical distribution of birds. *Ibis,* 1 (4): 449–54 [S52]. On page 453 he states: 'Celebes is in some respects peculiar, and distinct from both regions, and I am inclined to think it represents a very ancient land which may have been connected at distant intervals with both regions, or perhaps with some other continent forming a direct connexion with Africa'.

[303]  Annotation by Wallace: 'Birds can *make way* to land when out of *course,* other animals and plants depend only on wind and current'.

[304]  Hooker, J.D. 1859. *On the flora of Australia, its origin, affinities, and distribution; being an introductory essay to the Flora of Tasmania.* London: Lovell Reeve.

With every good wish. Believe me

Yours very sincerely

C. Darwin

Excuse this brief note, but I am far from well.—

# Amboyna

## 60. To John Gould, 30 September 1859

**Amboyna**

*Sept. 30th. 1859*

My dear Mr. Gould

I hasten to reply to yours of June 25th by return of post as requested, & it is pleasing to me to find that my collections excite so much interest.

[in margin] *Semioptera Wallacei G.R. Gray*

The following is an extract from my notebook:[305]

"It frequents the lower trees of the virgin forests, & is in almost constant motion. It flies from branch to branch, clings to the twigs, & even to the vertical smooth trunks almost as easily as a woodpecker. It continually utters a harsh croaking cry, something between that of *Paradisea apoda* and the more musical note of *P. regia*. The males at short intervals open & flutter their wings, erect the long shoulder feathers & expand the elegant shields on each side of the breast. Like the other Birds of Paradise, the females & young males far outnumber the fully plumaged birds, which makes it probable that it is only in the 2nd or 3rd year that the extraordinary accessory plumes are fully developed."

---

[305] Wallace quotes from *Notebook 4*, p. 135.

Fig. 22. John Gould

The bird seems to feed mostly on fruit, though it probably takes insects occasionally. The iris is of a deep olive;—the bill olive horny, the feet orange, and the claws horny. I have now obtained a few specimens of what appears to be the same bird from *Gilolo*,—but in these the crown is more decidedly *violet* & the plumes of the *breast* much larger.

I have now an interesting collection from *Menado* including the singular *Megacephalon* [Brush-turkey] and also one from Gilolo with I think a new & fine *Megapodius* [scrubfowl] & the superb *Pitta maxima* Forst. [Ivory-breasted Pitta], the finest bird I think of the whole family. After sending these off I am off to *Ceram* where I hope to do something good in ornithology. A paper lately sent to the *Linnaean* will give you my idea on the structure & origin of the Archipelago.[306]

Yours faithfully,

Alfred R. Wallace.

John *Gould* Esq.

---

[306]  Wallace, A.R. 1860. On the zoological geography of the Malay Archipelago. *Journal of the Proceedings of the Linnean Society: Zoology* 4: 172–84 [S53].

## 61. To Philip Lutley Sclater, 22 October 1859

Mr. Wallace's last communications are dated Amboyna, Oct. 22, 1859, whence he has sent us the valuable contributions to our pages which we have the pleasure of inserting in our present Number.[307] He further says,—

I have just packed up a large collection of Gilolo and Ternate birds, as well as those from Menado. The former are a much gayer lot, comprising a fine series of *Pitta maxima*, a new *Megapodius*, I think, handsomely banded on the back, and a *Semioptera*, which differs a little from the Batchian specimens in the much greater length of the breast plumes and other details. Is the *Calœnas* the true *nicobarica* [Nicobar pigeon]? If so, it is a unique case of a true land-bird ranging through the whole Archipelago, and beyond its limits from the Andamans to New Guinea. I do not know where Bonaparte got his information about its being arboreal. Here it is truly terrestrial, perching only to rest and sleep.

It is astonishing how little care even professed naturalists have given to determining localities. The localities of species given by the 'Dutch Scientific Commission' are full of errors.[308] *Ptilonopus monachus* and *P. hyogaster* are given to Celebes, whereas they are unknown there, but are abundant in Gilolo and Batchian; and exactly the same error is made with *Macropygia reinwardti*, which you will see in my collections, but not from Celebes. *Todiramphus funebris* is also unknown in Celebes, but common in Gilolo, so that the Dutch naturalists seem to have placed all their species of unknown locality in Celebes, acting as the French have done in giving to the little island of Vanikoro hundreds of insects which were never found there.

---

[307]  Wallace, A.R. 1860. The ornithology of Northern Celebes. *Ibis* 2 (6): 140–7 [S57].
[308]  The 'Natuurkundige Commissie voor Nederlandsch-Indië', active 1820–1850, was formed to explore the geology, botany, and zoology of the Dutch East Indies. Among its members were Salomon Müller (1826–1836), Ludwig Horner (1835–1838), and many others. Husson L.M. and Holthuis, L.B. 1955. The dates of publication of 'Verhandelingen over de natuurlijke Geschiedenis der Nederlandsche overzeesche Bezittingen'. Ed. C.J. Temminck. *Zoologische Mededelingen* 34: 17–24.

Among the other interesting species from Gilolo are a *Ptilonopus* and a *Platycercus*—both, I think, new; the beautiful *Ianthœnas halmaheira*, Bp., and several fine aquatic birds and Waders.

In a few days I commence work in Ceram, where I hope to make a very fine collection, especially of *Psittacidæ*, the Lories of Ceram surpassing even those of New Guinea in variety and beauty. I live in hopes too of a new *Semioptera*, or some equally interesting form.

I take every opportunity of purchasing live specimens of Parrots from the islands I may probably not visit, and hope to get most valuable materials for elucidating their distribution in the East, which is in the highest degree interesting. Between the *Lorius garrulus* of Gilolo and that of Batchian there is a constant difference in the size of the dorsal yellow patch: are they considered distinct species?

The species of Ceram birds mentioned in Bonaparte's 'Conspectus'[309] are very few: how is it, then, that it has such a name for fine birds? I know nothing fine from it, but the Lories, which are superb. However, I hope and believe it will produce some very fine things—new Pigeons, perhaps. The Cassowary is said to be abundant in Ceram, and to be the same as the New Guinea species. The *Tanysipteræ* are very puzzling: which is the true *T. dea*, Linn.? The Dorey and Ternate specimens seem almost identical, and in G. R. Gray's list, New Guinea specimens are put as *T. dea*.[310] If so, then the larger white-tailed species found in Amboyna and Ceram is undescribed, and is perhaps the same as the white-tailed specimens from the Kaisa Islands, sent with my Batchian collections.*

---

[309] Bonaparte, C.L.J.L. 1850. *Conspectus generum avium*, [T.I]. Lugduni Batavorum.

[310] Gray, J.E. and Gray, G.R. 1859. *Catalogue of the mammalia and birds of New Guinea, in the collection of the British Museum.* London: British Museum (Natural History).

The *Carpophaga perspicillata* of Amboyna differs also from those of Gilolo and Batchian in the much lighter colour of the head. Now, I believe in all these cases, where the difference is *constant*, we must call them distinct species. A 'permanent local variety' is an absurdity and a contradiction; and, if we once admit it, we make species a matter of pure opinion, and shut the door to all uniformity of nomenclature."[311]

## 62. From Charles Darwin, 13 November 1859

**Down Bromley Kent**

*Nov. 13./59*

My dear Sir

I have told Murray to send you by Post (if possible) a copy of my Book & I hope that you will receive it at nearly same time with this note.[312] (N.B I have got a bad finger which makes me write extra badly—). If you are so inclined, I sh$^d$ very much like to hear your general impression of the Book as you have

---

[311] Wallace here revisited the issues discussed in his 1858 paper, Note on the theory of permanent and geographical varieties. *Zoologist* 16: 5887–8 [S39]. The publication of this letter ends with a note by P.L. Sclater: 'Mr. Gray has named the Havre Dorey bird *T. galatea* (P.Z.S. [Proceedings of the Zoological Society] 1859, p. 154). That from Ternate must be the true *T. dea*, Ternate being the locality given for the *Alcedo dea* of the old authors. We believe that Mr. Gray refers the examples from Batchian and from the Kaisa islands to different species both undescribed'.

[312] This must refer to a list prepared by Darwin for the shipment of copies of the *Origin of species* at the time of publication; the book first appeared on 24 November 1859. Wallace received his copy in February 1860, see the remark in Darwin's letter of 18 May 1860 (CCD8: 556). Wallace's annotated copy survives today at Cambridge University Library, see Beddall, B.G. 1988.

thought so profoundly on subject & in so nearly same channel with myself. I hope there will be some little new to you, but I fear not much. Remember it is only an abstract & very much condensed. God knows what the public will think.

No one has read it, except Lyell, with whom I have had much correspondence. Hooker thinks him a complete convert; but he does not seem so in his letters to me; but he is evidently deeply interested in subject.—I do not think your share in the theory will be overlooked by the real judges as Hooker, Lyell, Asa Gray[313] &c.—

I have heard from Mr Sclater that your paper on Malay Arch. has been read at Linn. Soc, & that he was *extremely* much interested by it.[314]

I have not seen one naturalist for 6 or 9 months owing to the state of my health, & therefore I really have no news to tell you.—I am writing this at Ilkley Wells, where I have been with my family for the last six weeks & shall stay for some few weeks longer. As yet I have profited very little. God knows when I shall have strength for my bigger book.—

I sincerely hope that you keep your health; I suppose that you will be thinking of returning soon with your magnificent collection & still grander mental materials. You will be puzzled how to publish. The Royal Soc. fund will be worth your consideration.

With every good wish, pray believe me,

Yours very sincerely

Charles Darwin

I think that I told you before that Hooker is a complete convert. If I can convert Huxley I shall be content.—

---

[313] Asa Gray (1810–1888), American botanist and professor of natural history at Harvard University.

[314] The paper (S53) was read on 3 November 1859.

# Ceram

## 63. To Henry Walter Bates, 25 November 1859

**Ceram**

*Nov. 25th. 1859*

Dear Bates

Allow me to congratulate you ~~with~~ on your safe arrival home with all your treasures; a good fortune which I trust is this time reserved for me.[315] I hope you will write to me & tell me your projects. Stevens hinted at your undertaking a "Fauna of the Amazon Valley." It would be a noble work,—but one requiring years of labour, as of course you would wish to incorporate all existing materials & would have to spend months in *Berlin, Vienna* & *Milan* & *Paris* to study the collections of *Spix, Natterer, Oscolati, Castelnau* & *others*, as well as most of the chief private coll^ns of Europe.[316] I hope you may undertake it and bring it to a glorious conclusion.

I have long been contemplating such a work for this Archipelago,—but am convinced that the plan must be very limited to be capable of completion. It sh^d be little more than a *synopsis* on a uniform scale & plan,—*descriptions* in Latin only for brevity, but with copious introductions & remarks on *families* & *genera*. My idea also is that in such a work, it is the *duty* of an Entomologist to follow the arrangement of *Lacordaire* for the Coleoptera[317],—even where we differ from him,—as we sh^d thus so much facilitate reference & study, his work being in every ones hands. Excuse these remarks, but the subject interests me.—

I suppose you have now your collection pretty well arranged, & I shall be anxious to hear what is now the total amount of species in the chief families &

---

[315]   Henry Bates returned to England, via New York, in the summer of 1859.

[316]   Zoological collections by contemporary scientists, including Johann Baptist Ritter von Spix (1781–1826); Johann Natterer (1787–1843); Gaetano Osculati (1808–1894); and Francis de Castelnau (1810–1880).

[317]   Lacordaire, T. 1854–1876. *Histoire naturelle des insectes: genera des coléoptères.* Paris, 10 vols.

orders of insects.—You can tell me also what you think of my collections as far as you can judge in their *packed away* & perhaps dirty state. I now keep more duplicates than at first & hope you will reserve for me a set of your spare *Longicorns, Bupresti, Cicindelae, Cleri* & *Brenthi* & some types of Genera of Carabidae & *Curculios.*—Also in Butterflies, *Papilios,* & *Pieridae.* To these groups, I intend to confine my general collections.—

I have sent a paper lately to the *Linnean Soc.* which gives my views of the *principles of Geog. Distribution* in this Archipelago, of which I hope some day to work out the details.[318]

To judge of my coll$^n$ of other orders but Col. [Coleoptera] & Butt. [Butterflies] you must see *Mr Saunders* collections as he possesses my complete series, & the *Orthoptera, Hymenoptera* & *Diptera* must be now very extensive.

Your brother Frederick began a correspond$^{ce}$ with me, but does not go on. Please write soon.

I remain, dear Bates

yours very sincerely,

Alfred R. Wallace.

*H. W. Bates* Esq.

## 64. To Samuel Stevens, 26 November and 31 December 1859
### Awaiya, Ceram

*Nov. 26, 1859*

I have nothing particular to say now, except that Ceram is a *wretched place* for birds. I have been here a month, and have got *literally* not a single pretty or good bird of any kind, except the small Lory I sent before from Amboyna; and, what is

---

[318] Wallace, A.R. 1860. On the zoological geography of the Malay Archipelago. *Journal of the Proceedings of the Linnean Society: Zoology* 4: 172–84 [S53].

more, neither European residents nor natives know of a single handsome bird in the country, except one or two Lories and Pigeons, which I have not yet got or seen. When Mr. Gould and others talked about the very fine birds of Ceram, you should have asked them to specify them, that I might know what to inquire or look for.

My only hope is now in the eastern part of the island; but I cannot expect there anything but one or two fine Lories. In *Coleoptera* and Butterflies I shall do better, though almost all are the same as at Amboyna. I am at present confined to the house from the bites of an *Acarus* [mite], which produces inflamed sores on the legs, though it is invisibly small. My three best men have all left me—one sick, another gone home to his sick mother, and the third[319] and best is married in Ternate, and his wife would not let him go: he, however, remains working for me, and is going again to the eastern part of Gilolo.

**Passo, Island of Amboyna**

*Dec. 31, 1859*

My letter was returned to me because I had not prepaid the postage as far as Singapore. I now add a few lines. I have just arrived here, being quite tired of the barrenness of Ceram. I shall stay about three weeks, and then go to East Ceram and Kè, if possible.

# Amboyna

## 65. To Samuel Stevens, 14 February 1860

**Passo**

*Feb. 14, 1860*

I send you this *viâ* Marseilles, in order that you may get for me, as soon as possible, three cheap small double-barrelled guns, and send them overland to

---

[319] This was his Malay assistant Ali.

Ternate, to be ready for my next year's campaign to New Guinea. They are absolutely necessary for me, as I have now with me Charles Allen, who went out with me, and we must have a double quantity of tools to work with. He is now starting from N. Ceram and Mysole, while I go to E. Ceram and Kè. I expect to get some grand collections yet to send you. I am now packing up my Ceram and Amboyna collections to send you.

In birds they are miserably poor—only *one* being, I think, new, and very interesting from being a second species of Celebes' genus *Basilornis*. The few specimens of *Tanysiptera* were only obtained by two men going out for a month after nothing else; and the beautiful *Lorius domicella* was equally scarce, though domesticated specimens are abundant. There is scarcely anything else of interest but the unique *Platycercus amboinensis* (not found in Amboyna, however), which will show that my Dorey Bird was a distinct species.

## 65. From Daniel Hanbury, 17 February 1860

**London**

*17. Feb. 1860*

Dear Sir

Many thanks to you for the torch of "Rajah Dammar" [a resin] which Mr Stevens handed me yesterday. I have made a little slit in the palm-leaf in order to see the Dammar within and ascertain if it resembled any of that found in the London market. However the resin was too much broken to afford much information.

We much require information on Dammar, or rather as to what trees afford the particular sorts found in commerce. What, for instance, yields *Dammar batu*, or *Dammar puti* and what *Dammar Selan*? The latter is largely imported here.

I remain &c DH.

Mr. A.R. Wallace, care of Mr S. Stevens.

## 67. From Charles Darwin, 18 May 1860

**Down Bromley Kent**

*May 18th 1860*

My dear Mr Wallace

I received this morning your letter from Amboyna dated Feb. 16th, containing some remarks & your too high approbation of my book. Your letter has pleased me very much, & I most completely agree with you on the parts which are strongest & which are weakest. The imperfection of Geolog. Record is, as you say, the weakest of all; but yet I am pleased to find that there are almost more Geological converts than of pursuers of other branches of natural science. I may mention Lyell, Ramsay, Jukes, Rogers, Keyserling, all good men & true—Pictet of Geneva is not a convert, but is evidently staggered (as I think is Bronn of Heidelberg) & he has written a perfectly fair review in the Bib. Universelle of Geneva.—Old Bronn has translated my book, well-done also, into German & his well-known name will give it circulation.—I think geologists are more converted than simple naturalists because more accustomed to reasoning.[320]

Before telling you about progress of opinion on subject, you must let me say how I admire the generous manner in which you speak of my Book: most persons would in your position have felt some envy or jealousy. How nobly free you seem to be of this common failing of mankind.—But you speak far too modestly of yourself;—you would, if you had had my leisure done the work just as well, perhaps better, than I have done it.—

Talking of envy, you never read anything more envious & spiteful (with numerous misrepresentations) than Owen is in the Edinburgh Review.[321] I must

---

[320] Andrew Crombie Ramsay (1814–1891), Scottish geologist. Joseph Beete Jukes (1811–1869), geologist. Henry Darwin Rogers (1809–1866), geologist. Alexander Friederich Michael Leberecht Arthur von Keyserling (1815–1891), Russian palaeontologist. François Jules Pictet de la Rive (1809–1872), Swiss zoologist. Heinrich Georg Bronn (1800–1862), German palaeontologist and zoologist.

[321] [Owen, Richard]. 1860. [Review of *Origin* & other works.] *Edinburgh Review* 111: 487–532.

give one instance he throws doubts & sneers at my saying that the ovigerous frena of cirripedes have been converted into Branchiae, because I have not proved them to be Branchiae; whereas he himself admits, before I wrote, on cirripedes, without the least hesitation that these organs are Branchiae.—

The attacks have been heavy & incessant of late. Sedgwick & Prof. Clarke attacked me savagely at Cambridge Phil. Soc. but Henslow defended me well, though not a convert.[322]—Phillips has since attacked me in Lecture at Cambridge. Sir W. Jardine in Eding. New Phil. Journal.—Wollaston in Annal of Nat. History.— A. Murray before Royal Soc. of Edinburgh—Haughton at Geolog. Soc. of Dublin—Dawson in Canadian Nat. Magazine, and *many others*.[323]

But I am got case-hardened, & all these attacks will make me only more determinately fight. Agassiz[324] sends me personal civil messages but incessantly attacks me; but Asa Gray fights like a hero in defence.—Lyell keeps as firm as a tower, & this autumn will publish on Geological History of Man, & will there declare his conversion, which now is universally known.—I hope that you have received Hooker's splendid Essay.—So far is bigotry carried, that I can name 3 Botanists who will not even read Hooker's Essay!!

---

322  Adam Sedgwick (1785–1873), professor of geology, founded the Cambridge Philosophical Society in 1819. He read a paper 'on the succession of organic forms during long geological periods' at a meeting of the Society on 7 May 1860 (see Henslow to Hooker 10 May 1860, CCD8: 200). William Clark (1817–1865) was professor of anatomy at Cambridge University.

323  John Phillips (1800–1874), keeper of the Ashmolean Museum, Oxford gave the Rede lecture in Cambridge on 15 May 1860, see his *Life on the earth: its origin and succession* (London, 1860). William Jardine (1800–1874), a Scottish naturalist, published his review of the *Origin* in *Edinburgh New Philosophical Journal* n.s. 11: 280–9 (1860); T.V. Wollaston in *Annals and Magazine of Natural History* 5: 132–43 (1860); Andrew Murray (1812–1878), an advocate and naturalist, in the *Proceedings of the Royal Society of Edinburgh* 4: 274–91 (1860); Samuel Haughton (1821–1897), an Irish scientist, in his presidential address published in *Journal of the Geological Society of Dublin* 8: 137–56 (1860); and John William Dawson (1820–1899), a Canadian geologist, in the *Canadian Naturalist* 5: 100–20 (1860).

324  Jean Louis Rodolphe Agassiz (1807–1873), professor of natural history at Harvard University 1848–1873.

Here is a curious thing, a Mr. Pat. Matthew, a Scotchman, published in 1830 a work on Naval Timber & Arboriculture, & in appendix to this, he gives most clearly but very briefly in half-dozen paragraphs our view of natural selection. It is most complete case of anticipitation. He published extracts in G. Chronicle: I got Book, & have since published letter, acknowledging that I am fairly forestalled.[325]—

Yesterday I heard from Lyell that a German Dr Schaffhausen has sent him a pamphlet published some years ago, in which same view is nearly anticipated but I have not yet seen this pamphlet.—My Brother, who is very sagacious man, always said you will find that some one will have been before you.[326]—

I am at work at my larger work which I shall publish in separate volumes.— But from ill-health & swarms of letters, I get on very very slowly.—I hope that I shall not have wearied you with these details.—

With sincere thanks for your letter, & with most deeply-felt wishes for your success in science & in every way believe me,

Your sincere well-wisher

C. Darwin

---

[325] Patrick Matthew (1790–1874), Scottish farmer and author of *On naval timber and arboriculture* (London, 1831), summarized in Matthew, P. 1860. Nature's law of selection. *Gardeners' Chronicle and Agricultural Gazette* (7 April): 312–13.

[326] Hermann Joseph Schaaffhausen (1816–1893), a German anthropologist, wrote: Über Beständigkeit und Umwandlung der Arten. *Verhandlungen des Naturhistorischen Vereines der preussischen Rheinlande und Westphalens* 10: 420–51 (1853). Darwin's only brother was Erasmus Alvey Darwin (1804–1881).

# WAIGIOU, TERNATE, AND TIMOR
## 17 June 1860–5 July 1861

On 17 June 1860 Wallace set sail for the island of Waigiou near New Guinea, hoping to have more luck in his continued search for Birds of Paradise. His assistant Charles Allen, who was collecting independently on the nearby island of Mysol, badly needed supplies which Wallace planned to deliver. However, the winds carried Wallace's boat right past Mysol and it was impossible to turn back against them. Even worse, when they later stopped at a small island to make some rope, the anchor slipped and the boat was again pushed inevitably away by the winds. Two of Wallace's men were left stranded on the uninhabited island.

Two weeks later, on 4 July, Wallace and his team settled at the small village of Muka on the southern coast of Waigiou. He sent a boat to fetch his stranded men which failed to rescue them, but a second boat sent out was eventually successful. Although Wallace could hear the calls of Birds of Paradise near his temporary home, he could not acquire any. A month of collecting resulted in only two of the stunning Red Bird of Paradise. Local villagers told him that the birds were more easily procured on nearby Gam Island, so he proceeded in a small outrigger to the village of Bessir. Here at last he had more luck, so he stayed in a tiny hut for seven weeks during August and September 1860. He then had time to write to George Silk (1 September), praising Darwin's *Origin of species*. Wallace had now read it five or six times. He praised Darwin and the book in the highest possible terms: 'Mr. Darwin, has given the world a *new science*, & his name should, in my opinion, stand above that of every philosopher of ancient or modern times. The force of admiration can no further go!!!'

When his collection boxes were full, Wallace set out to return to Ternate on 2 October. The voyage should have taken only two weeks but a series of misfortunes and errors prolonged the voyage to 38 days. Finally, on 5 November 1860, he was back in his 'home' on Ternate, after an absence of about one and a half years. The voyage to Waigiou had produced 73 species of birds, 12 new to science, and 24 valuable specimens of the Red Bird of Paradise. A year of letters and mail were waiting for Wallace. Charles Allen had finally made it to Ternate as well and was waiting for Wallace with the spoils of his collecting trips on Ceram and Mysol. The birds were later described by George Gray at the British Museum.[327] Wallace almost immediately sent Allen on another trip to the northern part of Gilolo.

Wallace had to catch up with his mail. He wrote back to Darwin, and although only a fragment of the letter remains we agree with the editors of the *Correspondence of Charles Darwin* to date this to around December 1860.[328] Wallace had written comments on Von Buch's work on the Canaries while in Bessir in one of his notebooks and he essentially copied the paragraph into his letter to Darwin. The second part of the letter shows that Wallace had not yet realized the significance of what Henry Bates would later call insect mimicry. Wallace was puzzled how unrelated insects seemed to resemble one another in the same locality; perhaps, he suggested, the environment somehow determined their colouration. He also sent off another letter to Henry Bates, once again with high praise for the *Origin of species*.

After packing and shipping his latest consignment to Stevens, Wallace set out aboard a Dutch steamer for Timor on 2 January 1861, this time to the Portuguese port of Dilli. Ali and other assistants are not mentioned, so Wallace seems to have travelled alone. Wallace had a letter of introduction to Alfred Edward Hart, an English merchant captain and coffee planter. He invited Wallace to stay with him at his house at Malua about a mile outside the town in the foothills. Here Wallace met Cornish mining engineer Frederick F. Geach, who

---

[327]   Gray, G.R. 1861. Remarks on, and descriptions of, new species of birds lately sent by Mr. A.R. Wallace, from Waigiou, Misool and Gagie Islands. *Proceedings of the Zoological Society of London* (10 December): 427–38.

[328]   CCD8: 505.

remained his friend for many years. Darwin's book was again the topic of conversation and Hart borrowed Wallace's copy.

On 15 March Wallace wrote to his brother-in-law, Thomas Sims. As usual, the letter is full of advice and comments on best business practices. The tone and advice suggests that Wallace was looked up to as the great man of the family. At the end Wallace made some unusually candid remarks about his views of religion. It is the most detailed account we have of Wallace's views on religion, scepticism, and philosophy at this time.

Wallace next moved to the island of Bouru, just west of Amboyna, possibly because he had arranged to meet up once again with Charles Allen. His young English assistant was now a fully fledged collector and had made another trip to New Guinea, obviously to find additional specimens of Birds of Paradise. After a journey lasting from January to June 1861 to this inhospitable land, Allen finally landed in Bouru on 10 June to hand over his collections to Wallace and get instructions for the next destinations. Eleven days later, Wallace noted 'Charles left' in his notebook, maybe unaware at the time that they would never meet again.[329]

While collecting birds on Bouru, Wallace, according to his later account, first recognized what was to be called mimicry: 'I discovered in the island of Bouru two birds which I constantly mistook for each other, and which yet belonged to two distinct and somewhat distant families. One of these is a honeysucker named Tropidorhynchus bouruensis, and the other a kind of oriole, which has been called Mimeta bouruensis'.[330] Wallace was well satisfied that his collection of birds from Bouru included 66 species, of which 17 proved to be new to science.[331]

[329] Rookmaaker, K. and Wyhe, J. van. 2012. In Wallace's shadow: the forgotten assistant of Alfred Russel Wallace, Charles Allen. *Journal of the Malayan Branch of the Royal Asiatic Society* 85 (2): 17–54.

[330] MA2: 151. Also [Wallace, A.R.] 1867. Mimicry, and other protective resemblances among animals. *Westminster Review* (n.s.) 32 (173, 1 July): 1–43 [S134].

[331] Wallace, A.R. 1863. List of birds collected in the island of Bouru (one of the Moluccas), with descriptions of the new species. *Proceedings of the Zoological Society of London* (13 January): 18–36 [S72].

On 3 July 1861 Wallace boarded the Dutch mail steamer which took him back to Ternate. Here he made his final arrangements quickly and left Ternate and the Spice Islands, for the last time. He was again accompanied by Ali as they travelled on the steamer heading for Java.

# Waigiou

## 68. To George Silk, 1 September 1860 and 2 January 1861

**Bessir**

Sept. 1st. 1860

My dear George

It is now *ten months* since ~~I included a~~ the date of my last letter from England. You may fancy therefore that, in the expressive language of the trappers, I am half froze for news. No such thing. Except for my own private & personal affairs I care not a straw & scarcely ever give a thought as to what may be uppermost in the political world. In my situation old newspapers are just as good as new ones, & I enjoy the odd scraps, in which I do up my birds (advertisements & all), as much as you do your "Times" at breakfast. If I live however to return to Ternate in another month, I expect to find such a deluge of communications (including some arrears from you) that I shall probably have no time to answer any of them. I therefore bestow one of my solitary evenings on answering yours beforehand.

By the by, you do not yet know where I am, for I defy all the members of the Royal Geog. Soc. in full conclave to tell you where is the place from which I date this letter. I must inform you, therefore, that it is a village on the S.W. coast of the island of Waigiou, at the N.W. extremity of New Guinea.—How I came here would be too long to tell,—the details I send to my mother & refer you to her. While hon. members are shooting Partridges I am shooting or trying to shoot Birds of Paradise,—red at that, as Morris Haggar would say.[332]

But enough of this nonsense. I meant to write you of matters more worthy of a naturalists pen. I have been reading of late two books of the highest interest, but of most opposite characters, & I wish to recommend their perusal to you, if you have time for anything but work & politics. They are "Leon Dufour's Hist. of Prostitution"[333] & "Darwin's "Origin of Species." If there is an English translation of the first, pray get it. Every student of men & morals sh[d] read it, & if many who talk glibly of putting down the 'social evil' were first to devote a few days to its study, they would be both much better qualified to give an opinion on the subject & much more diffident of their capacity to deal with the question. The work is truly a history, & a great one, & reveals pictures of human nature more wild & incredible than the pen of the romancist ever dared to delineate. I doubt if many classical scholars have an idea of what were really the habits & daily life of the Romans as here delineated. Again I say, read it.

The other book you may have heard of & perhaps read, but it is not one perusal which will enable any man to appreciate it. I have read it through 5 or 6 times, each time with increasing admiration. It is the "*Principia*" of Natural History. It will live as long as the "Principia" of Newton. It shows that nature is, as I before remarked to you, a study that yields to none in grandeur & immensity. What are the cycles of Astronomy or even the periods of Geology to the vast depths of time contemplated in the endeavour to understand the slow growth of life upon the earth. The most intricate effects of the law of Gravitation,

---

[332] Morris Haggar may have been a childhood friend of Wallace and Silk. In 1841, the UK census lists a Morris Haggar born 1828 living in Hackney. A 'Mrs Haggar' is mentioned in a letter from John Wallace to M.A. Wallace, 12 October 1854 (NHM WP1/3/96/17).

[333] Dufour, P. 1851. *Histoire de la Prostitution chez tous les peuples du monde depuis l'antiquité la plus reculée jusqu'à nos jours*. Paris. There are notes on this in *Notebook 4*, p. 91. The book was never translated into English.

the mutual disturbances of all the bodies of the solar system, are simplicity itself, compared with the intricate relations & complicated struggle which has determined what forms of life shall exist & in what proportions. Mr. Darwin, has given the world a *new science*, & his name should, in my opinion, stand above that of every philosopher of ancient or modern times. The force of admiration can no further go!!!

### On board Steamer from Ternate to Timor
*Jan. 2nd. 1861*[334]

I have come home safe to Ternate & left it again. For two months I was stupified with my years letters, accounts, papers, magazines, & books, added to the manipulation, cleaning, arranging, comparing & packing for safe transmission to the other side of the world of about 16,000 specimens of Insects, Birds, & shells.

This has been intermingled with the troubles of preparing for new voyages, laying in stores, hiring men, paying or refusing to pay their debts, running after them when they try to run away,—going to the town with lists of articles *absolutely necessary* for the voyage, & finding *none* of them could be had for love or money,—conceiving impossible substitutes & not being able to get them either,—& all this coming suddenly upon me as my repose from the fatigues & miseries of an unusually dangerous & miserable voyage & you may imagine that I have not been in any great humour for letter writing.

I have however sent a little paper to the "Geographical"—on "The native trade with New Guinea."[335] I think I may promise you that in 18 months more

---

[334] This part of the letter dated 2 January 1861 is on a separate sheet and lacks all indications about the intended recipient. However, Wallace (ML1: 373) printed it as the continuation of his September letter to Silk.

[335] Wallace, A.R. 1862. On the trade of the Eastern Archipelago with New Guinea and its islands. *Journal of the Royal Geographical Society* 32: 127–37 [S65].

or less we may meet again if nothing unforeseen occurs, with wh. information I will conclude.

*ARW.*

# Ternate

## 69. To Charles Darwin, [December 1860]

From *"Von Buch on the Flora of the* Canaries."[336]

"On continents the individuals of one kind of plant disperse themselves very far, and by the difference of stations of nourishment & of soil produce *varieties* which at such a distance not being crossed by other varieties and so brought back to the primitive type, become at length permanent and distinct *species*. Then if by chance in other directions they meet with an other *variety* equally changed in its march, the two are become very *distinct* species and are no longer susceptible of intermixture."

P.S. "Natural Selection" explains *almost* everything in Nature, but there is one class of phenomena I cannot bring under it,—the repetition of the forms & colours of animals in distinct groups, but the two always occurring in the same country & generally on the *very same spot*. These are most striking in insects, & I am constantly meeting with fresh instances. Moths resemble butterflies of the same country—*Papilios* in the east resemble *Euploeas*, in America *Heliconias*. At

---

[336]   Buch, C.L. von. 1825. *Physicalische Beschreibung der Canarischen Inseln.* Berlin; translated in 1836 as *Description physique des Iles Canaries.* Paris. This might be Wallace's response to the remarks in Darwin's letter of 18 May 1860 regarding anticipations of the theory (CCD8: 219). Wallace copied these notes from his *Notebook 4*, p. 90 dating from c. August–September 1860, Bessir.

Amboyna I took on the same tree at the same time two longicorns of distinct genera, but so alike in colour & markings that I only separated them after some days—

Here also & at Macassar occurs together a *Malacoderm* [soft-winged beetle] & an *Elater* [click beetle] of exactly the same tints of metallic blue & soft orange & also similarly striate,—yet there is no affinity between them. A few days ago only I took a new & curious little *Cicindela* which so closely resembles in size & markings a *Therates* occurring with it, that I never know which it is till I take it out of my net;—yet there is no sign of a change in the structural characters which separate these *genera*. It seems to shew that colour, markings & texture of surface depend strictly on local conditions—Home Entomologists might do something in experiments on breeding insects, varying conditions of food light heat &c as much as they will bear. In domestic var$^s$. have you discovered what tends to produce white, black, or particular coloured variations? or what tends to produce spots rather than stripes.

*ARW.*

## 70. To Samuel Stevens, 7 December [1860]

**Ternate**

*December 7th*

I returned to Ternate a few days after the last mail had left here, having had a most hazardous voyage from Ceram and Waigiou. My collections are immense, but very poor, when it is considered that they are the result of nine months' collecting by two persons in East and North Ceram, Mysol, and Waigiou. Ceram is a wretched country; and the Papuan Islands, now that the cream is taken off by Aru and Dorey, are really not worth visiting, except for the Birds of Paradise.

My beetles, I am sorry to say, are most miserable—smaller and more obscure species than at Dorey, and only a few of the good ones found there, and none in any quantity.

In birds there is absolutely nothing good but the *Paradisea rubra* [Red Bird of Paradise], which is the only species that inhabits Waigiou, and is peculiar to that island.

I have been so busy with my mass of specimens (all wanting sorting and cleaning), and with my numerous letters and books (a whole year), that my mind has been too much unsettled to write. Next mail I shall write to all my entomological and ornithological friends who have been kind enough to send me communications.

I do not like the figure of *Semioptera wallacii* copied in 'The Ibis' from Gould's:[337] the neck-shields are not shown to advantage; and the white plumes should be raised much higher or laid down lower—they are neither one thing nor the other.

C. Allen starts in a week or two for N. Guinea—to the true locality for the rarer Birds of Paradise, and I trust he may be successful. The last voyage, with all its dangers and disappointments, has nearly sickened me, and I think in one year I shall return.

I seem to have all your letters but one (April 16, 1860).

## 71. To Philip Lutley Sclater, 10 December 1860

Mr. Wallace's letters from Ternate (of December 10th, 1860), enclosing the valuable paper already given (anteà, p. 283)[338], contain several passages which may interest our readers:—

I do not like the figure of *Semioptera wallacii*: the shoulder-plumes are not sufficiently erected; neither is the contrast of colour between the pure whiteness and the dark silky ash of the back sufficiently marked.

---

[337] Sclater, P.L. 1860. Note on Wallace's standard-wing, Semioptera wallacii. *Ibis* 2: 26–8, pl. II.

[338] Wallace, A.R. 1861. On the ornithology of Ceram and Waigiou. *Ibis* 3 (11): 283–91, pl. IX [S62].

The Dutch have just sent out a collector for the Leyden Museum to the Moluccas. He is now at Ternate, and goes to spend two years in Gilolo and Batchian, and then to N. Guinea. He will, of course (having four hunters constantly employed, and not being obliged to make his collecting pay expenses), do much more than I have been able to do; but I think I have got the cream of it all. His name is Bernstein;[339] he has resided long in Java, as doctor at a Sanatorium, and tells me he has already sent large collections to Leyden, including the nests and eggs of more than a hundred species of birds! Are these yet arranged and exhibited? They must form a most interesting collection.

Many thanks for your list of Parrots.[340] My collections already furnish many corrections of the localities. Allow me here to make a remark on the constant changes of specific names by yourself and Mr. Gray. It strikes me that, by forcing the law of priority to its extreme limits, you create a complicated synonymy, instead of settling it. Was not that law made to decide among several names already in use—not to introduce diversity where uniformity of nomenclature has hitherto existed? What is gained by changing *Eclectus linnæi* into *E. cardinalis*, and *Paradisea superba* into *P. atra*, when it is almost certain that such changes will not be generally adopted?

I believe the synonymy of Natural History will never be settled till a tribunal shall be appointed by general assent, from whose decrees there shall be no appeal. It matters absolutely nothing whether a bird has one name or another; but it is of the utmost importance that it should not have two or three at once. A synonymical catalogue, which should be authoritative and final by the general consent of naturalists in congress assembled, would be a work worthy of the

---

[339] Heinrich Agathon Bernstein (1828–1865) was a physician in a hospital at Gadok, near Buitenzorg (Bogor), Java, 1855–1860. He was appointed as the official government collector and worked in the Moluccas and New Guinea until 1865, with his headquarters in Ternate. He probably never met Wallace.

[340] Sclater, P.L. 1860. On the species of the genus Prioniturus, and on the geographical distribution of the Psittacidae in the Eastern Archipelago. *Proceedings of the Zoological Society of London* (13 March): 223–8.

century. Let ornithologists be the first in the field, and the other -ologists will soon follow.

The Cockatoos puzzle me greatly. You make my Lombock sp. *C. æquatorialis*, which Temminck says is peculiar to N. Gilolo and N. Celebes. Do you make it a synonym of *C. sulphurea*, which you do not mention? You will see small specimens of a Cockatoo from Mysol, which I thought were *C. æquatorialis*. I have just received a very small specimen from Gilolo, bearing the same relation to *C. cristata* that *C. sulphurea* does to *C. triton*. It will be, I suppose, quite new.

The larger and smaller specimens of *Megapodius* from Mysol are also curious. In colour they are exactly alike; but the size of the bill and feet is so different that they must be distinct. Between the *Trichoglossus* of Amboyna and Ceram and that of the Papuan Islands I can discover no difference, and I suspect that *T. nigrigularis* of G. R. Gray must be suppressed. You have left out *Lorius domicella* altogether from your list, giving *L. tricolor* to Amboyna in its place, which latter is wholly Papuan. *Eos cyanostriata* is a native of Timor-laut; and of *Eos reticulata* and *squamata* I saw nothing in Amboyna and Ceram, and believe they do not exist there. *Aprosmictus amboinensis* is a species strictly confined to Ceram, which you have not given. It is quite distinct from the *A. dorsalis* of New Guinea. The *Psittacidæ* of the Solomon Islands seem so exactly representative of those of New Guinea and the Moluccas, as to show that they must be included in the Papuan subregion, and (if true Lories are not found in New Caledonia) will mark its eastern limits. New Ireland and the eastern parts of New Guinea no doubt still contain many fine things in this group.

## 72. To Francis Polkingthorne Pascoe, 20 December 1860
### Ternate
*Dec' 20th. 1860*

My dear Mr. Pascoe

Yours of January & March last were only received by me on my return here last month. Many thanks for the papers you have sent me (the list of Longicorns of

Australia I have not however received) & especially for your ideas on the subject of publishing genera & species. Your plan has removed a considerable weight from my mind, for the quantity of minute & obscure species in my collections were beginning to frighten me.

My collections now just packed are immense as regards specimens (above 13,000) but they are particularly poor in Longicorns & in fact in Coleoptera generally. There will be no doubt some curious & new things among the small ones wh. are very numerous. They are from *Ceram, Mysol* & *Waigiou*, & if you are taking to Homoptera [sucking insects] you will find I think the largest collection of them you have ever seen.

The geographical distribution of Insects in the Archipelago is certainly far less strongly marked than that of *Birds* & *Mammals*, but I think that it may be in a great measure imputed to the much greater liability of insects to accidental dispersion. Still such cases as *Tmesisternus* & its allies strictly confined to the Australian region of the Archipelago & *Collyris* almost equally peculiar to the Indian region point to the same primary distribution in the one case as in the other. More generally however the same genera have I think subgroups with a characteristic "*facies*" in each region.

Insects moreover are much more affected ~~by~~ as to form by local circumstances than Birds & Mammals, thus the extreme general similarity in climate & vegetation between Borneo & N. Guinea has led also to a general similarity in the most prevalent insect forms, while the really close zoologic connection between Australia & N. Guinea is obscured by the very strongly contrasted physical features of the two countries being favourable to the propagation & development of distinct forms of insect life. Thus the *Anthribidae* [fungus weevils] of the Indian region have increased & developed rapidly in the equally luxuriant forests of the Moluccas & N. Guinea, while in the arid plains of Australia they barely exist.

On the other hand the numerous *Melolonthidae* [scarab beetles] of Australia favoured by the dry climate & numerous flowering shrubs of [word deleted] its plains, are unknown in the damp & flowerless forests of N. Guinea & the Moluccas, where also the more bulky forms of the Indian region have not been able to penetrate. Hence the excessive paucity of Lamellicornes in N. Guinea.

The distribution of insects is therefore more obscure & complicated than that of the vertebrates. It was originally the same, but it is more rapidly affected by time & the accidents of distribution & differences of climates.

In haste.

Yours very faithfully

Alfred R. Wallace

*F.P.Pascoe* Esq.

## 73. To Henry Walter Bates, 24 December 1860

**Ternate**

*Dec.ʳ 24th. 1860*

Dear Bates

Many thanks for your long & interesting letter. I have myself suffered much in the same way as you describe & I think more severely. The kind of "*tedium vitae*" you mention I also occasionally experience here. I impute it to a too monotonous existence.

I know not how or to whom to express fully my admiration of *Darwin's book* [*Origin of species*]. To him it would seem flattery, to others self-praise;—but I do honestly believe that with however much patience I had worked up & experimented on the subject, I could never have *approached* the completeness of his book,—its vast accumulation of evidence, its overwhelming argument, & its admirable tone & spirit. I really feel thankful that it has not been left to me to give the theory to the public.

Mr. Darwin has created a new science & a new Philosophy, & I believe that *never* has such a complete illustration of a new branch of human knowledge, been due to the labours & researches of a single man. Never have such vast masses of widely scattered & hitherto utterly disconnected facts been combined into a system, & brought to bear upon the establishment of such a grand & new & simple philosophy!

I am surprised at your joining to N. & S. banks of lower Amazon into one region. Did you not find a sufficiency of distinct sp[ecies]. at *Obydos* & *Barra* to separate them from *V. Nova* & *Santarem*?

239

I am now convinced that *insects* on the whole do not give such true indications in Zoological Geog. as Birds & Mammals because they have,—1st. such immensely greater chances of distribution, & 2nd. because they are so much more affected by local circumstances, also 3rd. because the sp. aren't to change quicker & therefore disguise a comparatively recent identity. Thus the insects of two originally distinct regions may rapidly become amalgamated,—or portions of the same region may come to be inhabited by very distinct insect faunas owing to differences of soil, climate &c. &c. This is strikingly shown here, where the insect fauna from Malacca to N. Guinea has a very large amount of characteristic uniformity, while Australia from its distinct climate & vegetation shows a wide difference. I am inclined to think therefore, that a preliminary study of first the *Mammals* & then the *Birds* are indispensable to a correct understanding of the geographical & physical changes on which the present insect distribution depends.

With regard to exchanges, I think it must be left till my return, which according to my present plans will not be delayed beyond a year & a half from this date. The groups I intend to collect generally are,—*Papilios* & *Pieridae* only among Lepidoptera;—& *Cicindelidae, Carabides* Lac. [Lacépède], *Buprestidae, Cleridae, Longicornes* & *Brenthidae* among Coleoptera.—Also illustrations of *genera* of Coleoptera generally & the more common of the remarkable or handsome species. If you will put by for me at your leisure the most complete sets you can spare of these groups, I shall (I have no doubt) be able to let you have an equal number of such specimens as you may desire.

In a day or two I leave for *Timor* where if I am lucky in finding a good locality, I expect some fine & interesting insects.

In haste

Yours faithfully,
Alfred R. Wallace.

H.W. *Bates* Esq.

# Timor

## 74. To Samuel Stevens, 6 February 1861

The last letters received by Mr. S. Stevens from Mr. Wallace are dated Delli in Timor, February 6th, 1861, and state that he had been there a month, and intended waiting two more. The country was barren, and, Australia-like, poor in insects; but birds were tolerably abundant, though not of very fine species.

## 75. To Thomas Sims, 15 March 1861

**Timor Delli**

*March 15th. 1861*

My dear Thomas

I will now try and write you a few lines in reply to your last 3 letters, which I have not before had time & inclination to do. First, about your *one eyed* & *two eyed* theory of art &c. &c. I do not altogether agree with you. We do not see *all objects* wider with two eyes than with one. A spherical or curved object we *do* see so, because our right & left eye each see a portion of the surface not seen by the other, but for that very reason the portion seen perfectly with both eyes is *less* than with one.

Thus we only see from a to a′ with both our eyes, the two side portions ab, a′ b′ being seen with but one eye & therefore (when we are using both eyes) being seen obscurely. But if we look at a flat object ~~on the~~ whether square or oblique to the line of vision we see it of exactly the same size with two eyes as with one because the one eye can see no part of it that the other does not see also. But in painting I believe that this difference of proportion where it does exist is far too small to *be given* by any artist and also too small to affect the picture if given.

Again I entirely deny that by *any means* the exact effect of a landscape with objects at various distances from the eye can be given on a flat surface; & more-

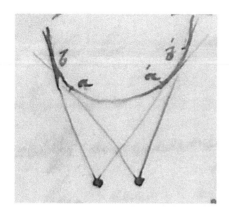

Fig. 23. Wallace's sketch in a letter to Sims, March 1861

over that the *monocular* clear outlined view is quite as true & good on the whole as the *binocular* hazy outlined view, & for this reason;— we cannot & do not see clearly or look at two objects at once, if at different distances from us. In a real view our eyes are directed successively at every object, which we then see clearly & with distinct outlines, everything else—nearer & farther being indistinct; but being able to change the focal angle of our two eyes & their angle of direction with great rapidity we are enabled to glance rapidly at each object in succession & thus obtain a general & detailed view of the whole. A house, a tree, a spire, the leaves of a shrub in the foreground, are each seen (*while we direct our eyes to them*) with perfect definition & sharpness of outline.

Now a monocular photo gives the clearness of outline & accuracy of definition & thus represents *every individual part of a landscape* just as we see it when looking at that part. Now I maintain that this is *right*, because no painting can represent an object both distinct & indistinct. The only question is, shall a painting show us objects as we see them when looking *at them*, or as we see them when looking at *something else* near them. The only approach painters can make to this varying effect of binocular vision & what they often do, is to give the most important & main feature of their painting *distinct* as we should see it when looking at it in nature, while all around has a subdued tone & haziness of outline like that produced by seeing the real objects when our vision is not absolutely directed to them.

But then if as in nature you turn your gaze to one of these objects in order to see it clearly you cannot do so, & this is a defect. Again, I believe that we actually see in a good photograph *better* than in nature, because the best camera

242

lenses are more perfectly adjusted than our eyes & give objects at varying distances with better definition. Thus on a picture we see at the same time near & distant objects easily & clearly, which in reality we cannot do. If we could do so everyone must acknowledge that our vision would be so much the more perfect & our appreciation of the beauties of nature more intense & complete; & insofar as a good landscape painting gives us this power it is better than nature itself;—& I think this may perhaps account for that excessive & entrancing beauty of a good landscape or of a good panorama.

You will think these ideas horribly heterodox, but if we all thought alike there would be nothing to write about & nothing to learn. I quite agree with you however as to artists using both eyes to paint & to see their paintings, but I think you quite mistake the do theory of looking through the "catalogue";—it is not because the picture can be seen better with one eye, but because its effect can be better seen when all lateral objects are hidden & the catalogue does this.— A double tube w$^d$ be better, but that cannot be extemporized so easily. Have you ever tried a *stereograph* taken with the camera only the distance apart of the eyes? That must give *nature*. When the angle is greater the views in the stereoscope show us, not nature, but a perfect *reduced model of nature* seen nearer the eye. It is curious that you sh$^d$ put *Turner* & the Prerhaph$^{tes}$ [Pre-Raphaelites] as *opposed* & representing binoc.[ular] and monoc.[ular] painting when Turner himself praises up the Prerhaph$^{tes}$ & calls Holman Hunt the greatest living painter!![341]

Now for your next letter wh. relates to your engraving processes. I have been too long out of the way of civilization to be able to judge of your probable success, but I cannot understand from your description how you are to get any detail in the dark parts of a picture, when the adhesion of powdered gum to a partially dried photo is the method employed unless you hit the moment when only the absolute black parts are dry & so retain no gum,—& this I sh$^d$ think

---

[341] Joseph Mallord William Turner (1775–1851) was a landscape painter. William Holman Hunt (1827–1910) was one of the founders of the Pre-Raphaelite Brotherhood. This group wanted to revitalize art through emphasis on detailed observation of the natural world.

must be very difficult. However the great point is does it bring out the detail & does it give the whole range of tone from high lights to deep shadows?

The points on which I long since asked information you say nothing about,— viz. are there any good lithographs or engravings yet published produced direct from photographs. Such are advertised by a German company & are alluded to in the Athenaeum as very successful. Now if yours are not likely to be *better*, it is not much use going on. You hint that Talbots & all other processes yet known give very poor results or are altogether impracticable. If such is the case & you believe you can do better I hope you will be able to get up a few good specimens against the next "Great Exhib$^n$". Surely in the slack season in Sept$^r$ you will have both *light* & *time*.

Now for your last letter of *June*. I am glad to find you are going on fairly in business. I met the other day a *photographer* who has been 8 years out in Australia & here.[342] There was too much competition in Australia & he came to Batavia & has travelled all through Java & a good part of Sumatra,—3 months at Macassar & is now going round the Moluccas staying 2 months at each place, He says it pays very well. The wealthy natives & chineese give a good price for Portraits of themselves, their wives & children & he takes good numbers of views for the stereoscope, all of wh. are sent home & readily sold.

As to the British Museum I am most strongly in favour of removing the Natural History collections simply because the Building in great Russel Street is totally *unadapted* for a museum of Nat. Hist. I made a sketch for a Natural Museum with a plan for its arrangement some time ago. It was shown to Prof. Owen & he said it was very good but for a national instition [institution] there was no occasion for the *economy* which I had insisted on among the other advantages of my plan.[343] If you like to see the paper ask Mr. Stevens to lend it

---

[342] Walter Bentley Woodbury (1834–1885) was co-owner of the Woodbury & Page studio in Batavia. They made many striking photographs of the people and places of the region. Four of their photographs were later engraved on wood as illustrations for Wallace's *Malay Archipelago*.

[343] A similar plan for the Natural History Museum by Wallace dated 1864 survives among his papers (NHM WP4/5).

to you. It is not therefore a question of what parts can be most easily moved. The building is *well fitted* for *Antiquities* & Books but *quite unfitted* for Zoological specimens.—

Now for Mr. Darwin's book. You quite misunderstand Mr. D.'s statement in the preface & his sentiments. I have of course been in correspondence with him since I first sent him my little essay. His conduct *has* been most liberal & disinterested. I think anyone who reads the Linn. Soc. papers & his book will see it. I *do* back him up in his whole round of conclusions & look upon him as the *Newton of Natural History*. You begin by criticizing the *title*. Now, though I consider the title admirable, I believe it is not Mr. Darwin's but the Publisher's, as you are no doubt aware that publishers *will* have a taking title, & authors must & do give way to them. Mr. D. gave me a *different* title before the book came out.[344]

Again you *misquote* & misunderstand *Huxley* who is a complete convert. Prof. Asa Gray & D^r Hooker, the two first botanists of Europe & America are converts. And *Lyell*, the first geologist living, who has all his life written against such conclusions as Darwin arrives at, is a convert & is about to declare or already has declared his conversion: A noble & almost unique example of a man yielding to conviction on a subject which he has taught as a master all his life, & confessing that he has all his life been wrong.

I see clearly that you have not yet sufficiently read the book to enable you to criticise it. It is a book in which every page & almost every line has a bearing on the main argument, & it is very difficult to bear in mind such a variety of facts, arguments and indications as are brought forward. It was only on the *5th* perusal that I fully appreciated the whole strength of the work, & as I had been long before familiar with the same subjects I cannot but think that persons less familiar with them, cannot have any clear idea of the accumulated argument by a single perusal. Your objections so far as I can see any thing definite in them are

---

[344]  See letter by Charles Darwin to Wallace, [13 July 1858].

so fully and clearly anticipated and answered in the book itself that it is perfectly useless my saying anything about them.

It seems to me however as clear as daylight that the principle of *Natural Selection must* act in nature. It is almost as necessary a truth as any of mathematics. Next,—the effects produced by this action *cannot be limited*. It cannot be shown that there *is* any limit to them in nature.—Again the millions of facts in the *numerical relations* of organic beings,—their *geographical distribution,*—their *relations of affinity*, the *modification of their parts* & organs, the *phenomena of intercrossing,—embryology* & *morphology,*—all are in accordance with his theory & almost all are necessary results from it;— while on the *other theory*, they are all isolated facts having no connection with each other & as utterly inexplicable & confusing as *fossils are* on the theory that they are special creations and are not the remains of animals that have once lived. It is this vast *chaos* of facts, which are explicable & fall into beautiful order on the one theory,—which are inexplicable & remain a chaos on the other, which I think must ultimately force Darwin's views on any & every reflecting mind.

Isolated difficulties & objections are nothing against this vast cumulative argument. The human mind cannot go on for ever accumulating facts which remain unconnected & without any mutual bearing & bound together by no law. The evidence for the production of the organic world by the simple laws of inheritance is exactly of the same nature as that for the production of the present surface of the earth, hills valleys plains rocks strata volcanoes, & all their fossil remains, by the slow and gradual action of natural causes now in operation. The mind that will ultimately reject Darwin must (to be consistent) reject Lyell also. The same arguments of *apparent stability* which are thought to disprove that organic species can change will also disprove any change in the inorganic world, & you must believe with ~~the~~ your forefathers that each hill & each river, each inland lake & continent, were created as they stand, with the *various strata* & their various fossils,—all appearances and arguments to the contrary notwithstanding.

I can only recommend you to read again Darwin's account of the *Horse* family & its comparison with *Pigeons*;—& if that does not convince or stagger you, then you are unconvertible. I do not expect Mr. Darwin's larger work will add any thing to the general strength of his argument. It will consist chiefly of the details (often numerical) of experiments & calculations of which he has already given the summaries & results. It will therefore be more confusing & less interesting to the general reader. It will prove to scientific men the accuracy of his details & point out the sources of his information, but as not one in a thousand readers will ever test these details & references the smaller work will remain for general purposes the best.

I am obliged to you & Mr. Fry[345] for your advice & fears about Mr. Stevens & beg to assure you that Mr. S. has my perfect confidence. He is a thorough business man, & sends me frequently full & detailed accounts, with balance sheets when the collections are finally disposed of. I have every reason to be *well satisfied* with all he has done for me. As he gets in the money, what can be spared is invested in the purchase of *E. India Railway* shares guaranteed at 5 per cent. These were recommended by Mr. *W.W. Saunders* of "*Lloyds*" as the best & safest investment & they are entered in the joint names of *Stevens & Saunders*. Mr. S. also sends me the Brokers accounts of the purchases.[346] So I think this is all right & better without the interference of lawyers.

I see that the Great Exhibition for 1862 seems determined on. If so it will be a great inducement to me to cut short the period of my banishment & get home in time to see it. I assure you I now feel at times very great longings for the peace & quiet of home,—very much weariness of this troublesome wearisome wandering life. I have lost some of that elasticity & freshness which made the overcoming difficulties a pleasure, & the country & people are now too

---

345  Probably Samuel Fry (1835–1890), a colleague photographer of Sims who lived in Brighton from 1857.

346  ML2: 360: 'My agent had invested the proceeds from time to time in Indian guaranteed railway stock, and a year after my return I found myself in possession of about £300 a year'.

Fig. 24. Wallace with Geach, in Singapore

familiar to me to retain any of the charms of novelty, which gild over so much that is really monotonous & disagreeable. My health too gives way & I cannot now put up so well with fatigue & privations as at first. All these causes will induce me to come home as soon as possible & I think I may promise if no accident happens to come back to dear & beautiful England in the summer of next year.

C. Allen will stay a year longer & complete the work which I shall not be able to do.[347]

I have been pretty comfortable here having for two months had the society of Mr. Geach, a Cornish mining engineer who has been looking for copper here.[348] He is a very intelligent & pleasant fellow, but has now left. Another Englishman, Capt. [Alfred Edward] Hart, is a resident here. He has a little house on the foot of the hills 2 miles out of town & I have a cottage (which was Mr. Geach's) a ¼ mile farther. He is what you may call a *speculative* man, has read a good deal, knows a little &

[347]  Charles Allen returned to Singapore a few weeks after Wallace's departure for England on 8 February 1862. He must have sent the last collections on to England. There is no indication that Allen afterwards continued to collect for Wallace in either Singapore or Malaysia.

[348]  Frederick F. Geach was working as a mining engineer in Timor when Wallace reached there in 1861. Geach moved to Singapore and worked in the region until 1866 employed by Paterson Simons & Co., the first British company to speculate in Malayan tin-mining on a large scale. The only photograph of Wallace during his voyage was taken in Singapore together with Geach in February 1862 (now kept in NHM WP2/1/3, and reproduced without Geach in Marchant, 1916, vol. 1: facing p. 36).

wants to know more & is fond of speculating on the most abstruse & unattainable points of science and philosophy. You would be astonished at the number of men among the captains & traders of these parts who have more than an average amount of literary & scientific taste;—whereas among the naval & military officers & various Government officials very few have any such taste,—but find their only amusements in card-playing & dissipation. Some of the most intelligent & best informed Dutchmen I have met with are trading captains or merchants.

This country much resembles Australia in its physical features, & is very barren compared with most of the other islands of the Archipelago. It is very rugged & mountainous, having no true forests, but a scanty vegetation of gum trees with a few thickets in moist places. It is consequently very poor in insects & in fact will hardly pay my expenses, but having once come here I may as well give it a fair trial. Birds are tolerably abundant, but with few exceptions very dull coloured. I really believe the *whole series* of Birds of the tropical island of Timor are less beautiful & bright coloured than those of Great Britain. In the mountains potatoes, cabbages & wheat are grown in abundance & we get excellent pure bread made by chinamen in Delli. Fowls, sheep, pigs & onions are also always to be had, so that it is the easiest country to live in I have yet met with, as in most other places one is always doubtful whether a dinner can be obtained.

I have been [on] a trip to the hills & staid ten days in the clouds, but it was very wet, being the wrong season. In about a month I leave here for Bouru where I spend 2 months to complete my Zoological survey of the Moluccas.

Having now paid you off my literary debts, I trust you will give me credit again for some long letters on things in general. Address now to care of *Hamilton, Gray and Co.*, Singapore, & with love and remembrances to all friends,

I remain

My dear Thomas

yours very faithfully,

Alfred R. Wallace.

*Mr. T. Sims.*

P.S. I see in a vol. of the Fam. Herald a statement that at the B. Ass. [British Association] at Glasgow 1855, Photographs *were exhibited* obtained by a coating of bitumen dissolved in sulp. ether applied to the stone. This is said to be *sensitive* to light & receives a picture from a negative. Sulp. ether dissolves away the bitumen when not acted on. How then washed & ready for the printer. They say the lithographer gives beauty. The same process on copper or steel produced etchings by coating the plate with gold which adheres only to the bare places. Here is a process quite diff. from the Bich. of potas. & gum that you mention, so that it seems you do not know all that has been done in England even so long ago as 1855.

I have been much interested by Claudet's description (at the last B. Ass. at Oxford) of the *Solar Camera* for enlarging photographs & his account of its theory. It is admirable & as he says will certainly produce great results. The only difficulty is the necessity of some equatorial motion to keep the suns image exactly on the centre of the lens. It will enable photographers to use the shortest lens in the field thus getting instantaneous pictures & then to enlarge these to any size preserving full strength & bringing out all detail. By such an arrangement there seems nothing that photography will not do.

Will you, next time you visit my mother, make me a little plan of her cottage showing the rooms & their dimensions, so that I may see if there will be room enough for me on my return? I shall want a good large room for my collections & as when I can decide exactly on my return it would be as well to get a little larger house beforehand if necessary. Please do not forget this.

Yours,

A.R.W.

P.S. You allude in your last letter to a subject I never touch upon because I know we cannot agree upon it. However, I will now say a few words, that you may know my opinions, & if you wish to convert me to your way of thinking, take more vigorous measures to effect it.

You intimate that the happiness to be enjoyed in a future state will depend upon & be a reward for, our *belief* in certain doctrines which you believe to constitute the essence of true religion. You must think therefore that belief is *voluntary* & also that it is *meritorious*. But I think that a little consideration will show you that *belief* is quite independent of our *will*, and our common expressions show it. We say, "I wish I could believe him innocent, but the evidence is too clear." Or,—"Whatever people may say, I can never believe he can do such a mean action." Now suppose in any similar case the evidence on both sides leads you to a certain *belief* or *disbelief*,—& then a *reward* is offered you for changing your opinion. Can you really change your opinion & belief, for the hope of reward or the fear of punishment?

Will you not say "As the matter stands I can't change my belief. You must give me proofs that I am wrong or show that the evidence I have heard is false & then I may change my belief." It may be that you do get more evidence & do change your belief, but this change is not *voluntary* on your part. It depends upon the *force of evidence* upon your individual mind, & the evidence remaining the same & your mental faculties remaining unimpaired, you *cannot* believe otherwise any more than you can fly.

Belief, then, is not voluntary,—how then can it be meritorious? When a jury try a case, all hear the same evidence, but 9 say *guilty* & 3 *not guilty*, according to the honest belief of each. Are either of these more worthy of reward on that account than the others? Certainly you will say, no! But suppose beforehand they all know or suspect that those who say *not-guilty* will be punished & the rest rewarded, what is likely to be the result? Why perhaps 6 will say *guilty* honestly believing it, & glad they can with a clear conscience escape punishment,—three will say *not-guilty* boldly, & rather bear the punishment than be false or dishonest,—the other 3 fearful of being convinced against their will, will carefully stop their ears while the witnesses for the defence are being examined, & delude themselves with the idea they give an honest verdict because they have heard only one side of the evidence. If any out of the dozen deserve punishment, you will surely agree with me it is these. Belief or disbelief is therefore not *meritorious*, & when founded on an unfair balance of evidence is blameable.

Now to apply these principles to my own case. In my early youth I heard, as 99-hundredths of the world do, only the evidence on one side, & became impressed with a veneration for religion which has left some traces even to this day. I have since heard & read much on both sides, & pondered much upon the matter in all its bearings. I spent, as you know, a year and a half in a clergyman's family & heard almost every Sunday the very best, most earnest & most impressive preacher it has ever been my fortune to meet with, but it produced no effect whatever on my mind.[349]

I have since wandered among men of many races & many religions. I have studied man & nature in all its aspects & I have sought after truth. In my solitude I have pondered much on the incomprehensible subjects of *space, eternity, life & death!* I think I have fairly heard & fairly weighed the evidence on both sides, & I remain an *utter disbeliever* in almost all that you consider the most sacred truths.

I will pass over as utterly contemptible the oft-repeated accusation that *sceptics* shut out evidence because they will not be governed by the morality of Christianity. You I know will not believe that in *my* case & *I* know its falsehood as a general rule. I only ask do you think I can change the self-formed convictions of 25 years? And can you think such a change would have anything in it to merit *reward* from *justice?*

I am thankful I can see much to admire in all religions. To the mass of mankind religion of some kind is a necessity. But whether there be a God & whatever be His nature; whether we have an immortal soul or not, or whatever may be our state after death, I can have no fear of having to suffer for the study of nature & the search for truth, or believe that those will be better off in a future state who have lived in the belief of doctrines inculcated from

---

[349] Wallace was hired as a master in the Collegiate School in Leicester, 1844–1845, where he lived in the house of the headmaster, Rev. Abraham Hill.

childhood, & which are to them rather a matter of blind faith than intelligent conviction.

*AR.W.*

This for yourself;—show the *letter* only to my mother.

P.S. *Write by next mail,* as circumstances have occurred which make it possible I may return home *this year. ARW.*

[Envelope]

*via Southampton.*[350]

Mrs. Sims

13a, Westbourne Grove

Bayswater

*London.*

---

[350]  Postmarks on this page: *[SOERABAYA]* FRANKO. Also: SINGAPORE P.O. *[illeg]* AUG 1861. And two postmarks: Paid X 19SP61 W LONDON.

# JAVA, SUMATRA, AND BACK HOME
## 6 July 1861–31 March 1862

Wallace reached Surabaya on the great island of Java on 16 July 1861.[351] He lodged in a hotel, for him a strange experience after so long staying in remote islands and jungle huts. During the month of August he visited the more mountainous regions south of the town. Of course, he was painfully aware that Java was the centre of the Dutch territories so the chances of finding zoological rarities on ground so heavily covered by previous naturalists were minimal. Again boarding a steamer, he made his way to Batavia, the capital of the Dutch East Indies, arriving on 18 September 1861. The Hotel des Indes provided him unparalleled luxury in a well-appointed room. Wallace opted for an excursion of a week to Buitenzorg (with a cooler climate), then reached in a coach drawn by four horses. Here he saw the magnificent palace of the Governor-General and the famous botanical gardens. Still looking for new collecting grounds, he continued inland and enjoyed exploring the different sceneries for about four weeks. He returned to Batavia on 28 October.

His next stop was the large island of Sumatra, much less developed than Java, where he hoped to collect some interesting insects, even if few would be new to science. Together with Ali, he embarked on the mail steamer *Prins van Oranje*, which had regular stops in Bangka and Riau on its way to Singapore. Disembarking in the small harbour of Muntok on 3 November 1861, he took a smaller boat across to the Sumatran mainland and then 100 miles upriver to the town of Palembang. Hoping to find favourable collecting spots, he travelled further inland until he came to the small hamlet of Moera Dua (Muara Dua), about 50 miles southwest of Palembang, where he stayed a fortnight.

---

[351]   The steamer *Ambon* arrived in Surabaya on 16 July 1860, *Javabode* 24 July 1860.

From this location Wallace wrote to Darwin on 30 November 1861. Wallace repeated his praise of *Origin of species* and told of three acquaintances in the East who had read the book. Baron Ferdinand von Richthofen, who had been reading the book while on the steamer with Wallace, had told him that it was easy for a geologist to accept the theory. Wallace also mentioned Darwin's 'great work'—the one Darwin had been working on when surprised by Wallace's Ternate essay in 1858. Only the first part of this large work (*Variation of animals and plants under domestication* of 1868) was published in his lifetime.

Not impressed with the collecting, Wallace pushed further inland and settled in a small village named Lobo Raman (Lubuk Raman) for the month of December 1861. Here he wrote his last letters in the East to Henry Bates and George Silk. The letter to Silk mentions Wallace's awareness of and interest in some of the key events that followed the publication of *Origin of species*—for instance, the collected *Essays and reviews* published by seven Church of England divines who promoted new findings in science and biblical scholarship that undermined traditional interpretations. Wallace was now ready to return home and start to sort his collections.

In early December Wallace 'received information' (letter to Sclater, 7 February 1862) that two live Birds of Paradise were for sale in Singapore. As a letter sent by mail was hardly likely to have followed him to his forest station, someone must have brought the news, at least as far as Palembang. Maybe his friend George Rappa forwarded the news, because he was listed as a passenger arriving back in Singapore with Wallace, but not listed as starting out with the steamer from Batavia.

Learning of this incredibly lucrative opportunity, Wallace made his way back to the coast and took the next mail steamer onwards to Singapore, where he arrived, together with Rappa, on 18 January 1862. The Zoological Society of London had offered Wallace a free first-class ticket back to England if he managed to bring these animals safely home. Wallace must have been relieved that the birds were still in good condition, and finally on 6 February he paid $400 for two birds to the commission agents Mark Moss and William Waterworth.

Wallace packed his bags for the final time. He gave money and equipment to his faithful companion Ali, and said goodbye to his friends Geach and Rappa. On

8 February 1862 the P&O steamer *Emeu* left Singapore, and after touching at Penang and Point de Galle reached Bombay 15 days later. A stop in Bombay gave Wallace the chance to obtain fresh vegetables and fruits for the precious birds. He then sailed on to Suez, crossed the Egyptian desert to Alexandria on the new rail line, and finally reached the island of Malta on 17 March 1862. Immediately he cabled Philip Sclater, the secretary of the Zoo, to ask for instructions, because he had hoped that a zookeeper would have come to meet him and take charge of the precious cargo. The telegraph office must have been quite unfamiliar with Birds of Paradise as they transcribed the message as 'garadisi bards'!

Speed was essential, so Wallace took the ferry to Marseilles and then a train across France to Boulogne. Wallace landed at Folkestone on 31 March 1862. He stayed at the large waterfront Pavilion Hotel where he wrote to Sclater about the safe arrival of the two Birds of Paradise. His great voyage had come to an end.

# Java

## 76. To Mary Ann Wallace, 20 July 1861

**Sourabaya, Java**

*July 20th. 1861*

My dear Mother

I am as you will see now commencing my retreat westwards, & have left the wild & savage Moluccas & New Guinea for Java the garden of the East, & probably without any exception the finest island in the world. My plans are to visit the interior & collect till *November*, & then work my way to Singapore so as to return home & arrive in the spring.

Travelling here will be a much pleasanter business than in any other country I have visited, as there are good roads, regular posting stages & regular inns or lodging-houses all over the interior, & I shall no more be obliged to carry about with me that miscellaneous lot of household furniture,—bed, blankets, pots kettles & frying pan,—plates, dishes & wash-basin, coffeepots & coffee, tea, sugar & butter,—salt, pickles, rice, bread & wine,—pepper & curry powder, & half a hundred more odds & ends, the constant looking after which, packing & repacking, calculating & contriving,—have been the ~~constant~~ standing plague of my life for the last 7 years. You will better understand this when I tell you that I have made in that time about *80 movements* averaging *one a month*, at every one of which all of these articles have had to be rearranged & repacked by myself according to the length of the trip, besides a constant personal supervision to prevent waste or destruction of stores in places where it is impossible to supply them.

Fanny wrote me last month to know about how I should like to live on my return. Of course, my dear mother, I sh$^d$ not think of living anywhere but with you, after such a long absence,—if you feel yourself equal to housekeeping for us both;—& I have always understood that your cottage would be large enough. The accommodation I should require is, besides a small bedroom, one large room, or a small one if there is, besides a kind of lumber room where I could keep my cases & do rough & dirty work. I expect soon from Thomas a sketch plan of your cottage, by which I can at once tell if it will do,—If not, I must leave you & Fanny to arrange as you like about a new residence. I should prefer being a little way out of town in a quiet neighbourhood & with a garden, but near an *omnibus route*, & if necessary I could lodge at any time for a week in London. This, I think, will be better & much cheaper than living close to Town, & rents anywhere in the West End are sure now to rise owing to the approaching great Exhibition. I must of course study economy, as the little money I have made will not be all got in for a year or two after my return.

As I came here the steamer from the Moluccas passed us & my letters & I believe the box from Mr Stevens with my clothes &c ~~passed~~ went on to Ternate & I shall have to wait a month for them. I have not had a letter from you for *six months*, but am glad to hear from Fanny that you are quite well as when I heard of the dreadfully severe winter I was rather alarmed.

You must remember to write to me by the *middle of November mail*, as that is probably the last letter I can receive from you.

I send this letter to Fanny, who will most likely call on you & talk over matters. I am a little confused arriving in a new place with a great deal to do & living in a noisy hotel,[352] so different to my usual solitary life,—so that I can not well collect my ideas to write any more, but must remain,

My dear Mother

Your ever affectionate Son,

Alfred R. Wallace.

*Mrs. Wallace*
*Sourabaya, Java July, 1861*[353]

## 77. From François de Caumont LaPorte, 26 August 1861

Singapore

*26th August 1861*

Dear Sir,

I received with much pleasure your letter of July 2—Ternate / 20—Surabaya[354]—the invoice in return for your box must be since a long time in the hands of Mr Stevens (London).

I am getting for you a *complete* set of my duplicates from Siam, Cambodia and the Malayan Peninsula, from the last I have certainly many sorts new for you. Since two years my people collecting in Tigor, Tringanu &c (I have 3000 sorts from the Peninsula)—the box will contain 1200 specimens.

---

352  Possibly Hotel der Nederlanden, Surabaya, Java.
353  Added in pencil, in another handwriting: 'to Mrs Wallace'. And again on separate part of page, in pencil: 'for Mrs. Wallace'.
354  Wallace must have started to write the letter on 2 July 1861 when he was on his way from Bouru to Ternate, finishing it on 20 July after arriving in Surabaya, Java.

I desire in exchange a complete set of your duplicates from Dorey, Ternate, Buru and Timor, from those localities I have nothing—from Menado I am almost in the same case. I have a great many from Ceram; my people are since one year in Sumatra and they are going to Borneo. I have people in Cambodia and Birmania [Burma] and I am going to send an expedition to Laos very soon.

You will find the box ready for you at Singapore. If I wait not there it would be in the hands of the owner of the hotel d'Europe M<sup>r</sup> Castelyns.³⁵⁵

I should like you on your arrival in England to make me an invoice of 2000 insects that will be an advance of 800 that I will pay ~~by~~ with insects of Birmania, Laos, Cambodia and Labuan when they will come.

I remain, Dear Sir, in expecting the pleasure of seeing you here

Yours Sinc.<sup>ly</sup>

Count de Castelnau

Consul of France at Siam

hotel d'Europe, Singapore

## 78. To Philip Lutley Sclater, [11–15] and 20 September 1861

### Sourabaya

*Sept<sup>r</sup>. 1861*

My dear D<sup>r</sup> Sclater,³⁵⁶

---

³⁵⁵ The Frenchman J. Casteleyns started a modest establishment in Singapore called the Hotel d'Europe at Hill Street in 1857, moving to Beach Road in 1860. The hotel moved again in 1865 when Casteleyns occupied the Hotel de l'Esperance buildings at the Esplanade and renamed it the Hotel d'Europe. Castelyns managed the hotel until 1869.

³⁵⁶ Wallace was in Surabaya from 11–15 September 1860 before catching the steamer to Batavia on the 15th (arriving on the 17th). In the extract of this letter printed in *Ibis* 4 (1861): 95–6 (S66), Sclater states that 'Mr. Wallace's last letters are dated from Batavia, Sept. 20th'.

"Better late than never" is a much to be admired proverb, & in accordance with its precept I now beg to thank you for the honour of my election to the "British Ornithologists Union", I really don't know how long ago.[357]

My excuses for this long forgetfulness are twofold:—1st. I received a printed notice of my election as Hon. Member of the *B.O.U.* without the slightest clue as to what the *B.O.U.* was. It was only some months afterwards a receipt of the 4th no. of the "Ibis" that I discovered the mystery.—2nd. For the last four or five years I have received my correspondence in lumps,—six to 12 months at a time, on my return from some land to which the beneficent agency of the Post Office does not extend. I have at such times generally had so much to do, in cleaning sorting studying & packing my collections,—& so much anxiety & trouble in preparing at the same time for a fresh voyage, hunting up men, fitting up boats, laying in stores &c &c that I have often been obliged to defer the perusal of my home news till again under weigh [way] or till again located in some barbarous solitude.

Having begun a letter I will give you a few notes about *Bouru*, where I staid two months after leaving Timor. From the existence of *Babirusa* in the island I had been somewhat doubtful whether its fauna would not prove more *Celebesian* than *Moluccan*. I was soon however satisfied that it is a true Moluccan island, though a very poor one. Most of the common Amboyna & Ceram forms occurred, some absolutely identical, others sufficiently modified to be characterized as distinct species. The [birds] *Tanygnathus, Polychlorus, Eclectus, Geoffroyus, Eos* & *Trichoglossus* as well as the *Aprosmictus* occur as in Ceram, the *Tanygnathus* being the only one which varies from the type, wanting the black markings in the wings. *Lorius* is altogether absent, as well as *Corvus, Buceros* & *Cacatua* genera which are present in every other island from Celebes eastwards. This deficiency does not rest alone on the fact of my not having met with them, though that would be pretty good proof, they being all ubiquitous & noisy

---

[357] Wallace was elected as an honorary member (not being resident in England) of the British Ornithologists Union on 9 June 1860.

birds,—but on the universal testimony of the natives, many of whom know all these birds from their visits to other islands but are quite sure their own country is destitute of them.

The flycatchers (3-4 sp.) seem new as well as a very common *Mimeta* near *M. forsteri* of Ceram, & a *Tropidorhynchus* I suppose the *T. buruensis* Q & G. [Quoy & Gaimard] though in "Bonaparte's Conspectus" that species is given to *Celebes* where I never found the genus. The pigeons are mostly known species except a fine *Treron* with very brilliant yellow marked wings, & I heard of other species of the same group occasionally met with. A single specimen of *Tanysiptera* seems different from the *Ceram* sp., & a *Pitta* near *macklotti* & *celebensis* but sufficiently distinct, is also unique. I was much surprised to find, besides the *Ptilonopus viridis* of Amboyna, the beautiful *P. prasinorrhous* G.R.G. [Gray] which I had just discovered in *Ké*, then found in *Goram*, afterwards in *Waigiou*, & I think there can be no doubt it is also found in *Ceram*, but birds seem so thinly scattered over that large island that it would take years to acquire a proper knowledge of its ornithology.

At *Bouru* I shot a *Glareola* the first time I have met with the genus. I found Coleoptera & grubs in its stomach. Its sternum shows it to be a true wader though a most curious & abnormal form.

The Cassowary is absent from Bouru & from every Moluccan island except Ceram, yet I had been positively assured it was common in Bouru. The error has arisen thus. The people of the little island of Bouru at the W. end of Ceram often get young *Cassowaries* from the main island to bring up. The traders of Cajeli in Bouru buy these & then take them to Amboyna for sale, often in company with young *Babirusas*. This happened when I was there. Of course the Amboyna merchants purchasing these animals from Bouru residents & having no reason for hunting up their pedigree, take it for granted that *Cassowaries & Babirusas* are found wild in Bouru.

I will give another example. In the Papuan island of Mysol the Cassowary does not exist, but last year the Rajah possessed a live *Ceram Cassowary* which he had brought from Wahai in Ceram. Now Mysol is known among the Bugis traders as a district of *Papua*, & if one of these men had bought & taken this bird to Singapore or Macassar he would declare it came from Papua. It thus becomes most difficult to ascertain the true locality of the *Psittaci* & *Paradiseus* which are articles of trade, & are carried about by the traders from place to place often passing through several hands before they are sold to a European. In this manner *Lorius domicella* has come into our list as a Papuan bird. Lesson I have no doubt bought a specimen at Dorey as you may often buy Celebes or Ternate birds in any of the trading ports of New Guinea. So *Cacatua citrinoc-ristata* which you have given to Timor in your list, on the authority I suppose of the Paris Museum, does not belong there, as you will see by my specimens which are the same as those from Lombock called by you *C. aequatorialis*. Here in Java I have purchased a specimen of *C. citrinocristata* but can get no indication of its locality. I suspect however it may be *Timor laut* to which the beautiful *Eos cyanostriata* undoubtedly belongs, and prahaws from Macassar to that island often touch at Timor. *Eos reticulata* & *squamata* I do not know, but they certainly do not belong to Amboyna or Ceram. In fact Muller, Forsten & the other naturalists of the Dutch Scientific Commission[358] seem to have been most careless in investigating localities, or even in noting accurately the species actually shot in the several islands, so that the authority of the Leyden Museum can not de be depended upon for locality in any doubtful cases.

**Batavia**

*Sept. 20*

I have just received yours of July 22. I know Rozenberg[359] well: he is a most lucky fellow. In one trip of a few weeks on N.W. coast of N. Guinea he fell in with lots of rare things wh. it took me years to find. He showed me the head of the Cassowary. My assistant, Allen visited the same spot & staid 3 days but could not get it & in three months in Salwatty & the main land could never get

---

[358] The 'Natuurkundige Commissie voor Nederlandsch-Indië'. Wallace mentions Salomon Müller (1826–1836) and Eltio Alegondas Forsten (1811–1843), who travelled in Celebes and the Moluccas between 1840 and 1842.

[359] Carl Benjamin Hermann von Rosenberg.

a specimen. Young ones are however often brought to the Moluccas & Macassar & it must some day come to Europe. I am still doubtful about the *Aru species*, but it must surely be one of the new sp. lately described, as the young birds are brought thence annually *by dozens*! In Allen's last coll[n] are *many rarities* & some novelties, wh. I hope to have the pleasure of showing you myself next spring. He is now at Sula (*Sulla bessi* on the maps), an island of the Celebes group from which I expect many new things. I have *seen* a new *Loriculus* from it.[360]

I have also just received G.R. Gray's list of Molucca birds.[361] He makes plenty of n.s. [new species] but I think is only perpetuating error by admitting in any form such birds as *Buceros lunatus, Megalaima australis, Goura coronata* & a host of others. I note upwards of 20 wh. are palpable errors either of locality or identification.

Please do not print this gossiping letter, except an extract or two.

With best wishes I remain

Yours very faithfully<br>Alfred R. Wallace

P.L. Sclater, M.A.

## 79. To Frances Sims, 10 October 1861

### In the Mountains of Java

*Oct. 10th. 1861*

My dear Fanny

I have just received your 2nd. letter in praise of your *new house*. As I have said my say about it in my last I shall now send you a few lines on other subjects.

---

[360]  Annotation by Sclater in the left margin written vertically: 'I never heard of this new Chalcopsitta & doubt it'.

[361]  Gray, G.R. 1860. List of birds collected by Mr. Wallace at the Molucca Islands, with descriptions of new species, etc. *Proceedings of the Zoological Society of London* 28 (26 June): 341–66.

I have been staying here a fortnight *4000* feet above the sea in a fine cool climate, but it is unfortunately dreadfully wet & cloudy.

I have just returned from a three days excursion to one of the great Java volcanoes 10,000 feet high. I slept two nights in a house 7500 ft. above the sea. It was bitterly cold at night as the hut was merely of plaited bamboo, like a *sieve*, so that the wind came in on all sides. I had flannel jackets & blankets & still was cold, & my poor men with nothing but their usual thin cotton clothes passed miserable nights lying on a mat on the ground round the fire which could only warm one side at a time.

The highest peak is an extinct volcano with the crater nearly filled up, forming merely a saucer on the top, in which is a good house built by the government for the old Dutch naturalists who surveyed & explored the mountain. There are a lot of strawberries planted there, wh. do very well, but there were not many ripe. The common weeds & plants of the top were very like English ones, such as *buttercups, sow-thistle, plantain, wormwood, chickweed, charlock, St. John's wort, violets* & many others, all closely allied to our common plants of those names but of distinct species. There was also a *honeysuckle* & a tall & very pretty kind of *cowslip*. None of these are found in the low tropical lands & most of them only on the tops of these high mountains. Mr. Darwin supposed them to have come there during a glacial or very cold period, when they could have spread over the tropics & as the heat increased, gradually rose up the mountains. They were as you may imagine most interesting to me, & I am very glad that I have ascended *one* lofty mountain in the tropics, though I had miserable wet weather & had no view, owing to constant clouds & mist.

I also visited a semi-active volcano close by continually sending out steam with a noise like a blast-furnace—quite enough to give me a conception of all other descriptions of *volcanoes*.

The lower parts of the mountains of Java, from 3000 to 6000 ft have the most beautiful tropical vegetation I have ever seen. Abundance of splendid tree ferns, some 50 feet high, & some hundreds of varieties of other ferns,—beautiful leaved plants as *Begonias, Melastomas* & many others & more flowers than are generally seen in the tropics. In fact this region exhibits all the beauty the tropics can produce, but still I consider & will always maintain that our own *meadows & woods & mountains* are *more beautiful.* Our own weeds & wayside flowers are far prettier & more varied than those of the tropics. It is only the *great leaves* & the *curious looking* plants, & the deep gloom of the forests, & the mass of tangled vegetation, that astonishes & delights Europeans, & it is certainly grand & interesting & in a certain sense beautiful;—but not the calm, sweet, warm beauty of our own fields,—& there is none of the brightness of our own flowers; a field of buttercups, a hill of gorse, or of heath; a bank of foxgloves & a hedge of wild roses and purple vetches surpass in *beauty* anything I have ever seen in the tropics. This is a favourite subject with me but I cannot go into it now.

Send the accompanying note to *Mr. Stevens immediately.*[362] You will see what I say to him about my collections here. Java is the richest of all the islands in Birds but they are as well known as those of Europe & it is almost impossible to get a new one. However I am adding fine specimens to my collection, which will be altogether the finest known of the birds of the Archipelago, except perhaps that of the Leyden Museum, who have had naturalists collecting for them in all the chief islands for many years with unlimited means.

Give my kind love to Mother to whom I will write next time

Your affectionate Brother

Alfred R. Wallace.

*Mrs. Sims.*

---

[362] The note to Stevens, presumably also dated October 1861, has not been located.

# Sumatra

## 80. To Charles Darwin, 30 November 1861

### Sumatra, 100 miles E. of Bencoolen

*Nov.ʳ 30th. 1861*

My dear Mr. Darwin

On an evening in the wet season in these central forests of Sumatra, I occupy myself in writing a few lines to you to say a few things which I may otherwise forget altogether. About a year & a half back I wrote to you from somewhere Eastward of Ceram, with more digested remarks on your book, but the letter with one to my agent Mr. Stevens never I believe reached Amboyna. All I can remember of them now is to the effect that repeated perusals had made the whole clearer & more connected to me & the general & particular arguments clearer & more forcible than at first.

I have lent the book to two persons here in the East, neither of them with any but the vaguest & most general knowledge of or taste for Natural History but both men of much reading & with a taste for speculation & theory wh. as Bentham[363] says is but another term for thought. The first Mr Duivenbode[364] of Ternate, a Dutchman educated in England, read it three times through before he returned it to me, expressing himself so much pleased & interested that he wished to master the whole argument. The other a merchant captain[365] settled at Timor Delli with whom I lived 3 months kept it all the time, was constantly reading it & we made it a subject of conversation almost whenever we met, & when I was leaving he did not return it till the steamer arrived going over the recapitulations of the chapters & the conclusion to get the most of it he possibly could.

---

[363]  Jeremy Bentham (1748–1832), philosopher.
[364]  Maarten Dick van Renesse van Duivenbode (1805–1878) was a Dutch business-man on Ternate. His estate was a 45 minute walk north of Ternate town.
[365]  Alfred Edward Hart, resident of Timor (MA1: 295).

These humble testimonies prove I think both the attractive manner in which you have treated the subject & the clearness with which you have stated & enforced the arguments; & I trust they will be as pleasing to you as I ~~trust~~ assure you they were to me.

I met the other day on board the Dutch steamer the Geologist accompanying a Prussian voyage of discovery &c. *"Baron Richthofen"*.[366] He has been through Java where he has made large collections of fossils—He [is] going now to leave the ship & travel in British India & afterwards go to the Amoor & overland to Europe to study & improve himself in Geology. He seems a very intelligent & good naturalist. He was reading your Book wh. he had borrowed in Java, & on my asking him if he was a convert,—he smiled & said "It is very easy for a Geologist."

I have also seen *Dr. Schneider*[367] who has geologized in Timor—He assures me he found many teeth of *Mastodon*,—also *terebratulae orthoceras* & other molluscs. He had given most of his collection to the Zoologist of the same Prussian Expedition but says he has published descriptions of all the Timor fossils in the Batavian Journal of Nat. Sciences, I forget the Dutch title but you know the work no doubt—They are to be published this month I think he told me.

I trust your great work goes on & is soon to appear. I hope however you will have it copiously illustrated. I am sure it will be for the publishers interest to do so as it will I have no doubt double the circulation. There are so many things that are weak when merely mentioned or described by ~~figures~~ numbers which become clear & strong when an appeal is made to the eye. The *varieties of*

---

[366] Baron Ferdinand von Richthofen (1833–1905), a German geologist visiting Asia as a member of the Eulenberg Expedition of the Prussian Government between 1860 and 1862. He travelled from Batavia to Singapore on the steamer *Prins van Oranje* on 1 November 1861, where Wallace was also listed as a passenger (*Javabode*, 2 November 1861).

[367] Carl Friedrich Adolph Schneider (b. 1821), army health officer in the Dutch East Indies, 1852–1866. Wallace refers to a paper on Timor fossils: Schneider, C.F.A. 1863. Bijdrage tot de geologische kennis van Timor. *Natuurkundig Tijdschrift van Nederlandsch Indië* 25: 87–107.

*pigeons*, the *stripes of horses*, the variations in *ants*, the formation of *honeycombs* & a hundred other things will all be better for good illustrations—They will also make the book newer & more distinct from its forerunner in the eyes of the public to whom it will otherwise appear as perhaps a mere enlargement. If this point is not decided, pray take it seriously into consideration.

I see nothing but the Athenaeum so know little of what is going on among Naturalists. Huxley & Owen seem to be at open war, but I cannot glean that any one has ventured to attack you fairly on the *whole question*, or ventured to answer the whole of your argument—I see by his advertisement D[r] Brees professes to do so but have no notice of his book.[368]

I have lately been very unsuccessful owing to unprecedented wet weather (the effect I suppose of the comet). This is my [remainder of page excised]

P.S. The shabby way in which your opponents try [some words excised] is amusing. First comes Owen[369] with his new interpretation of what naturalists mean by "*creation*" which it turns out is not *creation* at all, but "the *unknown manner in which species have come into existence*"!!! Phillips[370] I see adopts this new interpretation which ought certainly to raise up the bench of bishops against them, for what then becomes of "*special creation*" & "*special adaptation*" & "intelligent forethought visible in each special creation",—if *creation* is not creation at all, but a mere convenient expression of ignorance of the laws by which species have originated.

---

[368] Huxley and Owen's debate about the differences between the brains of apes and humans was carried out in letters in the *Athenaeum* in March and April 1861. Charles Robert Bree (1811–1886), physician and zoologist. Wallace refers to Bree, C.R. 1860. *Species not transmutable, nor the result of secondary causes: being a critical examination of Mr Darwin's work entitled 'Origin and variation of species'*. London and Edinburgh.

[369] Wallace paraphrases Owen's 'axiom of *the continuous operation of the ordained becoming of living things*' in [Owen, Richard]. 1860. [Review of the Origin of Species.] *Edinburgh Review* 111: 487–532, on p. 500.

[370] John Phillips (1800–1874), professor of geology at the University of Oxford. Phillips, J. 1860. *Life on earth, its origin and succession*. Cambridge and London.

Has my friend Bates sent you his papers on Amazon Insects, in wh. he gives details of "variation" which are a most valuable contribution to our cause. His papers also shew what mere blind work is the making of species from what may be called *chance collections*;—a species here, another there, a ♂ from one locality a ♀ from another, here a rare variety, & from another place the typical form. Of course with such materials & with localities often incorrectly indicated each succeeding elucidator has but added to the confusion; & the whole has formed a chaos, till some one collecting & observing for years over an extensive but connected district has been able to bring the whole into order.[371]

## 81. To Henry Walter Bates, 10 December 1861
### Sumatra (Lobo Raman, 100 miles E. of Bencoolen)
*December 10th. 1861*

My dear Bates

I should have written to you before to thank you for your paper on the "Papilios",[372] but I somehow never can post up my correspondence till I get into some savage wilderness like that in which I am at present. I have here read your paper with great attention & also with great pleasure, and I trust it is but the first of a long series which will establish your own fame and at the same time demonstrate the simplicity & beauty of the Darwinian philosophy.

Your paper is in every respect an admirable one, & uncontestably proves the necessity of minute & exact observation over a wide extent of country to enable a man to grapple with the more difficult groups, unravel their synonymy & mark out the limits of the several species & varieties. All this you have done

---

[371] Part of the paper on which this letter is written has been excised. There is an annotation in the lower left-hand corner of the second page: 'A. Wallace, Nov.20 / 1861'; and of the last page: '[Anon] Sumatra'.

[372] Bates, H.W. 1860. Contributions to an insect fauna of the Amazon valley. Diurnal Lepidoptera. [Read 5 March and 24 November 1860.] *Transactions of the Entomological Society of London* n.s. 5 (1858-61): 223–8, 335–61.

& have besides established a very interesting fact in Zoological Geography, that of the southern bank of the river having received its fauna from Guyana, & not from Brazil.

There is however another fact I think of equal interest & importance which you have barely touched upon, & yet I think your own materials in this very paper establish it, viz. that the river in a great many cases limits the range of species or of well-marked varieties. This fact I considered was shown, by the imperfect materials I brought home, to obtain both to the Amazon & Rio Negro. I read a paper before the Zool. Soc. on the Monkeys of the Lower Amazon & Rio Negro in which I stated that in almost every case for 800 miles up the R. Negro, the species were different on the opposite banks of the river.[373] Guyana species come up to the E. bank, Columbian sp. to the west bank, & I observed that it was therefore important that travelers collecting on the banks of large rivers sh^d note from wh. side every specimen came. Upon this D^r Gray came down upon me with a regular floorer. 'Why,' said he, 'we have specimens collected by Mr. Wallace himself marked Rio Negro only.' —

I don't think I answered him properly at the time, that those specimens were sent from near Barra before I had the slightest idea myself that the species were different on the opp. banks. In Mammals the fact was not so much to be wondered at, but few persons would credit that it would extend to Birds & winged insects. Yet I am convinced it does, & I only regret that I had not collected & studied birds there with the same assiduity I have here, as I am sure they would furnish some most interesting results.

However, "revenons à nos papillons" [to return to butterflies]. It seems to me that a person with no special knowledge of the district would have no idea from your paper that the species did not in almost every ~~case~~ instance occur on both banks of the river. In only one case do you specially mention a species being found *only* on the N. bank (*Ergeteles*). In other cases, except where the insect is

---

[373] Wallace, A.R. 1852. On the monkeys of the Amazon. *Proceedings of the Zoological Society of London* 20 (14 December): 107–10 [S8].

local & confined to one small district, no one can tell whether they occur on one or both banks. *Obydos* you only mention once, Barra & the Tunantins not at all. I think a list of the sp. or var. occurring on the S. bank or N. bank only sh^d have been given, and would be of much interest as establishing the fact that large rivers do act as *limits* in determining the range of species. From the localities you give it appears that of the 16 sp. & sub. sp. [of Papilio] peculiar to Amazon, 14 occur only on the S. bank. Also, that of the *Guyana* sp. *all* pass to the S. bank.

These facts I have picked out. They do not appear. It would seem therefore that Gu[y]ana forms having once crossed the river have a great tendency to become modified, & then never recross. Why the Brazilian species should not first have taken possession of their own side of the river is the mystery. I sh^d be inclined to think that the river bed is comparatively new, & that the S. plains were once continuous with Guyana,—in fact, that Guyana is older than N. Brazil, & after it had pushed out its alluvial plains into what is now N. Brazil an elevation on the Brazilian side made the river cut a new channel to the northward, leaving the Guyana sp. isolated, exposed to competition with a new set of species & thus led to them becoming modified, as we now find them.

The phenomenon of a tract of country having been peopled from one now separated from it & not from that of wh. it forms a part, is too extraordinary not to require some special & extraordinary cause, & the one I have mentioned seems capable of producing the effects, & by no means improbable (however unexpected) in itself. The whole district is I fear too little known geologically to test the supposition. The N. mountains of Brazil however *are* of recent elevation, since *fishes* of the chalk period are found at great ~~elevations~~ heights. This would bring their ~~elev~~ upheaval into the tertiary period & it may have continued to a recent period. Now if there are no proofs of such recent upheaval in the S. Guyana mountains, the theory thus far receives support.

I regret that your time was not divided more equally between the N. & S. banks, but I suppose you found the S. so much more productive in new & fine things. I suppose you will turn now to the *Coleoptera* & give us the

*Cicindelidae* on the same plan, & I hope you have made arrangement for a lot of copies, each part paged consecutively to form complete separate works when finished.

I am here making what I intend to be my last collections, but am doing very little in insects, as it is the wet season & all seems dead. I find in those districts where the seasons are strongly contrasted the good collecting time is very limited, only about a month or two at beginning of dry & a few weeks at commencement of rains. It is now two years since I have been able to get any beetles owing to bad localities & bad weather, so I am getting disgusted. When I do get a good place it is generally very good but they are dreadfully scarce. In Java I had to go 40 miles inland in the E. part & 60 miles in W. to get to a bit of forest, & then I got scarcely anything.

Here I have had to come 100 miles inland (by Palembang) & even here in the very centre of E. ~~forest~~ Sumatra, the forest is only in patches, & it is the height of the rains, so I get nothing,—a longicorn is a rarity, & I suppose I shall not get as many species in 2 months as I have done in 4 days in a good place. I am getting however some sweet little *Lycaenidae* [gossamer-winged butterflies], which is the only thing that keeps me in spirits.

I hope to be home before the opening of the *Exhibition* & look forward to seeing you in London though I fear my collection will be in dreadful confusion till towards the winter. I think my priv. coll. of Col. [Coleoptera] & Lep. [Lepidoptera] will be probably more extensive in *specimens* than yours, as I have a complete series from every island & chief locality (which amount to about *30*), and as I intend to re-ticket, catalogue & arrange them all, as well as my extensive collection of birds, I shall have work for years, a labour of love to wh. I look forward with much pleasure.

Remember me kindly to your brother Frederick who I also hope to see, & to have the pleasure of showing him a few of my Eastern gems.

Wishing you health & strength to make known your rich collection & careful observations to the world (a task which I soon trust to be myself labouring), I remain, yours very sincerely,

Alfred R. Wallace.

H.W. Bates Esq.

I should not wonder but your paper will convert Hewitson.[374] He is not I think very susceptible to general arguments, but this will come home to his very bosom & touch his feelings if anything will. I hope you have sent Darwin a copy I am sure it would please him.

P.S. I quite agree with you as to the affinity of the *Crassus* group with *Ornithoptera*,—a note to the same effect has stood in my "Boisduval"[375] for years. I doubt however the propriety of placing [Papilio] *Dolicaon* &c with *Protesilaus*. I am now anxious to compare the Eastern forest group *Polydorus* &c. with *Oeneas*, to see if there is affinity of structure.

## 82. To George Silk, 22 December 1861 and 20 January 1862
### Lobo Raman, Sumatra
*Dec.* 22nd. 1861

My dear George

Between eight & nine years ago, when we were concocting that absurd book, "Travels on the Amazon & Rio Negro,"[376] you gave me this identical piece of paper, with sundry others,—& now having scribbled away my last sheet of '*hot pressed writing*,' & being just 60 miles from another, I send you back your gift, with interest,—so you see a good action, sooner or later, find its sure reward.

I now take my pen to write to you *a letter*, I hope for the *last time*;—for I trust our future letters may be *viva voce*, as an Irishman might remark, & our

---

374   William Chapman Hewitson (1806–1878).

375   Boisduval, J.B.A.D. ed. 1832–1835. *Voyage de l'Astrolabe. Faune entomologique de l'Océanie*, 2 vols.

376   Wallace, A.R. 1853. *Narrative of travels on the Amazon and Rio Negro*. London: Reeve & Co [S714].

epistolary correspondence confined to *notes*. In fact, I really do think & believe I am coming home, & as I am quite uncertain when I may be able to send you this letter, I may possibly not be very long after it. Some fine morning (before the Exhibition opens) I expect to walk into 79, Pall Mall;[377] & I suppose I shall there find things in general much about the same as if I had walked out yesterday & come in tomorrow. There will you be seated on the same chair, at the same table, surrounded by the same account books, & writing upon paper of the same size & colour as when I last beheld you.

I shall find your inkstand, pens, & pencils in the same place, & in the same beautiful order, which my idiosyncrasy compels me to admire, but forbids me to imitate. (Could you see the table at which I now write, your hair would stand on end at the reckless confusion it exhibits!) I suppose you have now added a few more secretaryships to your former multifarious duties. I suppose that you still come every morning from Kensington & return there in the evening, & that things at the Archdeacon's go on precisely & identically as they did eight years ago.

I feel inclined to parody the words of Cicero & ask indignantly, "How long, O Georgius, will you thus abuse our patience? How long will this sublime indifference last?"[378] But I fear the stern despot *habit* has too strongly riveted your chains, & as after preliminary years of torture the indian fanatic can at last sleep only on his bed of spikes,—so perhaps now, you would hardly care to change that daily routine, which has lasted so long, even if the opportunity should be thrust upon you. Excuse me, my dear George, if I express myself too strongly on this subject which is truly no business of mine, but I cannot see without

---

[377]   Apparently, 79 Pall Mall was Silk's office address. The same address was in this period used by John Sinclair, whom he served as secretary, as well as by the Society for the Propagation of the Gospel in Foreign Lands.

[378]   Cicero, Oratio in Catalinam (Catiline Orations), I.I: 'Quo usque tandem abutere, Catilina, patientia nostra? Quam diu etiam furor iste tuus nos eludet? Quem ad finem sese effrenata iactabit audacia?'—'How long, O Catiline, will you abuse our patience? And for how long will that madness of yours mock us? To what end will your unbridled audacity hurl itself?'

regret my earliest friend devote himself so entirely, mind & body, to the service of others.

It is an age since you wrote to me last & yet you might have found plenty to write about without touching upon Politics. "Essays & Reviews" & "the Gorilla war" might have filled a page & you might have told me whether my last paper on *"New Guinea native trade"* was read at the Geographical, or any notice taken of it.[379]—Did you go to see *"Blondin"*, have you heard *Mr. Fechter*, have you read *"Great Expectations"*.[380] On all these famous matters a line or to from you would have been acceptable, whereas even my last somewhat lengthy epistle has not elicited a word. But I must excuse you;—writing is too much your daily toil,—we will make up for it all when I return & I will talk with you & argue with you on every subject under the sun—except *party politics*.

I am here in one of the places unknown to the Royal Geog. Soc. situated in the very centre of E. Sumatra, 100 miles from the sea all round. It is the height of the wet season, & pours down strong & steady, generally all night & half the day. Bad times for me, but I walk out regularly 3 or 4 hours every day, picking up what I can, & generally getting some little new or rare or beautiful thing to reward me. This is the land of the two horned Rhinoceros, the Elephant, the tiger, & the tapir; but they all make themselves very scarce, & beyond their tracks & their dung, & once hearing a rhinoceros *bark* not far off, I am not aware of their existence.

This, too, is the very land of monkeys;—they swarm about the villages & plantations,—long-tailed & short-tailed, & no tail at all,—white, black, &

[379] Powell, B. 1860. *Essays and reviews.* London: Parker & Son. The 'gorilla war' was a controversy about claims of the American explorer Paul du Chaillu (1835–1903) about the ferocity of gorillas. Wallace's paper: On the trade of the Eastern Archipelago with New Guinea and its islands. *Journal of the Royal Geographical Society* 32: 127–37 (S65), was read on 13 January 1862.

[380] Jean François Gravelet-Blondin (1824–1897) was a famous French tightrope walker who first appeared on stage in London in 1861. Charles Albert Fechter (1824–1879) was an Anglo-French actor then appearing at the Princess's Theatre, London. Charles Dickens' *Great Expectations* was first published in 1861.

gray; they are eternally racing about the tree-tops, & gamboling in the most amusing manner. The way they do jump is *"a caution to snakes"*! They throw themselves recklessly through the air, apparently sure, with one of their four hands, to catch hold of something. I estimated one jump by a long-tailed white monkey, at 30 feet horizontal, & 60 feet vertical from a high tree on to a lower one; he fell through, however, so great was his impetus, on to a still lower branch, & then without a moments stop, scampered away from tree to tree, evidently quite pleased with his pluck. When you startle a lot & one takes a leap like this, it is amusing to watch the others, some afraid & hesitating on the brink till at last they pluck up courage, take a run at it, & often roll over in the air with their desperate efforts. Then there are the long-armed apes, who never *walk* or *run*, but travel altogether by their long arms, swinging themselves along from bough to bough in the easiest & most graceful manner possible.

But I must leave the monkeys & turn to the men, who will more interest you, though there is nothing very remarkable in them. They are Malays speaking a curious half unintelligible Malay dialect,—Mahometans but retaining many pagan customs & superstitions. They are very ignorant, very lazy, & live almost absolutely on *rice* alone, thriving upon it however just as the Irish do or did on potatoes. They were a bad lot a few years ago, but the Dutch have brought them into order by their admirable system of supervision & government. By-the-by, I hope you have read *Mr. Money's* book on *"Java"*.[381] It is well worth while, & you will see ~~how~~ that I had come to the same conclusions as to Dutch colonial government from what I saw in Menado.

Nothing is worse & more absurd than the sneering prejudiced tone in which almost all English writers speak of the Dutch Government in the East.—It never has been worse than ours has been, & it is *now* much better; & what is greatly to their credit & not generally known, they take nearly the same pains to establish order & good government in those islands & possessions wh. are an

---

[381]  Money, J.W.B. 1861. *Java; or how to manage a colony.* London: Hurst and Blackett.

annual loss to them, as in those which yield them a revenue. I am convinced
their system is *right* & ours *wrong* in principle,—though of course in the practical working there may be & must be defects, & among the dutch themselves, both in Europe & India, there is a strong party against the present system, but they are mostly merchants & planters, who want to get the trade & commerce of the country made free, wh. in my opinion would be an act of suicidal madness, & would moreover *injure* instead of benefiting the natives.

Personally, I do not much like the dutch out here, or the dutch officials;—but I cannot help bearing witness to the excellence of their government of native races, gentle yet firm, respecting their manners, customs, & prejudices, yet introducing everywhere European law, order & industry.

### Singapore

#### — *January 20th. 1862*

Thus far had I written when I received yours of Nov.14th. It really pained me to find you so desponding & surprised me to hear that you are still burthened with the support of relations who I had imagined would by this time have been able even to take upon themselves the burthen you have so long borne.

Your brothers can surely now support themselves, & I have I think heard that your sister has musical talents & skill to enable her to support herself;—& I had certainly imagined that she was somewhat of a strong minded young lady who would have scorned vulgar prejudices & disregarded your wishes on this matter:—and I must say I should have admired her conduct more had she done so.

On the question of marriage, we probably differ much. I believe a good wife to be the greatest blessing a man can enjoy & the only road to happiness but the qualifications I should look for are probably not such as would satisfy you. My opinions have changed much on this point. I now look at intellectual companionship as quite a secondary matter, & should my good stars ever send me an affectionate good tempered & domestic wife, I should care not one iota for accomplishments or even for education.

I cannot write more now. I do not yet know how long I shall be here, perhaps a month. Then ho! for England!

In haste

Yours most affectionately,

Alfred R. Wallace.

*G. Silk Esq.*

———

# Singapore

## 83. From Mark Moss, 3 February 1862

Dear Sir

Having consulted with M^r Waterworth[382] concerning the Paradize Birds, which you wish to purchase, I may mention that we are inclined to sell at $500 cash. If these terms suit you, you can take them.

Yours Obediently,

M. Moss.

*3 February 1862*

---

[382] In the *Straits calendar and directory* for 1862, Mark Moss and William Waterworth are listed among the principal inhabitants, both with the address: Mount Vernon, Siffken Road. Mark Moss is listed among the 'merchants and agents' and as an auctioneer in Raffles Place (pp. 69, 72). William Waterworth was the Singapore partner of William and George Waterworth, established 1854, in Raffles Place (p. 71). Mark Moss arrived from England in 1835.

## 84. From Mark Moss, 6 February 1862

Received from A. R. Wallace Esq$^r$ the sum of $400 four hundred dollars for two living Paradize Birds.

M Moss

**Singapore**

*6 February 1862*

---

## 85. To Philip Lutley Sclater, 7 February 1862

**Singapore**

*Feb. 7th. 1862*

To *P.L. Sclater Esq.*, Secretary to the *Zoological Society of London*.

Dear Sir

About 2 months back I received information, when in the interior of Sumatra, that there were two live *Paradise Birds* in Singapore. I immediately determined to come & enquire about them & accordingly proceeded to this place a full month earlier than I had intended. They were in the hands of a European merchant who was well aware of their value & asked an exorbitant price. As however they seemed in excellent health, had been in Singapore 3 months & in possession of a Bugis trader a year before that, I determined if possible to obtain them.

After protracted negotiations I have purchased them for $400 (nearly £100), & *tomorrow* I take them on board the steamer for Europe.

I was afraid to let it be known that a free passage had been offered me with them, as the demands would then have risen proportionally. I therefore obtained a *promise to sell* before applying to the Manager of the P & O Company, when I was much surprised & disappointed to find that *no order for a free passage* had been sent out, but merely instructions to *take care* of the Birds if sent on board.

Under these circumstances I was at first inclined to give up any idea of completing the purchase, but on mature consideration I thought that you would

279

certainly blame me for want of confidence in the offers made me by yourself, & the promise of a free passage repeatedly made me through *Mr. Sam. Stevens.* I have had no official notice of my terms having been accepted by the Zool. Soc. except that in one of your letters you say "I suppose Stevens has told you that the Zoological Society has accepted your terms for Birds of Paradise"; but I presume from this that a resolution to that effect exists in the Books of the Society.

My intention is to stay at *Malta* & send a *Telegram* asking instructions, but post this letter here, to show that the Birds are purchased for the Zoological Society.

I remain Dear Sir,

Yours very faithfully,

Alfred R. Wallace.

# Ceylon

## 86. To Samuel Stevens, 16 February 1862

### Galle, Ceylon

*Feb. 16th. 1862*

Dear Stevens[383]

I posted a letter at Singapore with an enclosure to Mr. Sclater informing you of my having purchased *2 living Paradisea papuanus* in Singapore, & of my *not* having got a *free passage* with them.

They have reached here in perfect health, & if they pass the ordeal of the *dry atmosphere* of the Red Sea & Egypt, I shall have great hopes of success. I go now viâ *Bombay* as the steamers by that route are less crowded. The price is the same & I have 3 or 4 days in Bombay & can lay in a stock of fruit for the birds.

---

[383]  Annotated: '*2nd letter*'.

I am somewhat doubtful about how to bring them home as it will I fear be now stormy—cold in the channel. I calculate to reach *Malta* about March & stay there a week as I am informed it is very mild or even warm there at that time. The next steamer would leave Malta about March *25th* which would bring me to Southampton about April 4th or 5th.

I do not want to delay my return longer than that, & if Mr Sclater thinks it not safe to bring them home so early I should much prefer his sending one of his *keepers* to receive them from me either at *Malta* or in the S. of France, (say *Avignon*) as I am informed Marseilles is sometimes very cold & stormy in March.

My plan therefore is as follows: You will receive my letters about March 15. I arrive at Malta about March *17th* & immediately Telegraph to Mr. Sclater *if the birds are alive*, & I wish him immediately to Telegraph instructions to me, whether to bring them on myself by next steamer to Southampton,—or whether he will send a man to receive them at Malta or at any place on the railway line in the S. of France.

The steamer however remains only 6 hours in Malta, so I must stay there & cannot leave again till March *25th* & reach Marseille *27th*.

Notwithstanding the letter of instructions I have to look after the birds myself or they get neither food water or cleaning regularly. Such things want a servant to themselves, & I fear that if we have bad weather & I am sick, they will suffer.

I much regret not being able to bring the *Siamang* [gibbon], which was a most interesting little creature. It is however going round the Cape & if well attended to may live.

Communicate this immediately to Mr. Sclater.

In haste —

Yours very faithfully

Alfred R. Wallace

*Sam¹. Stevens Esq.*

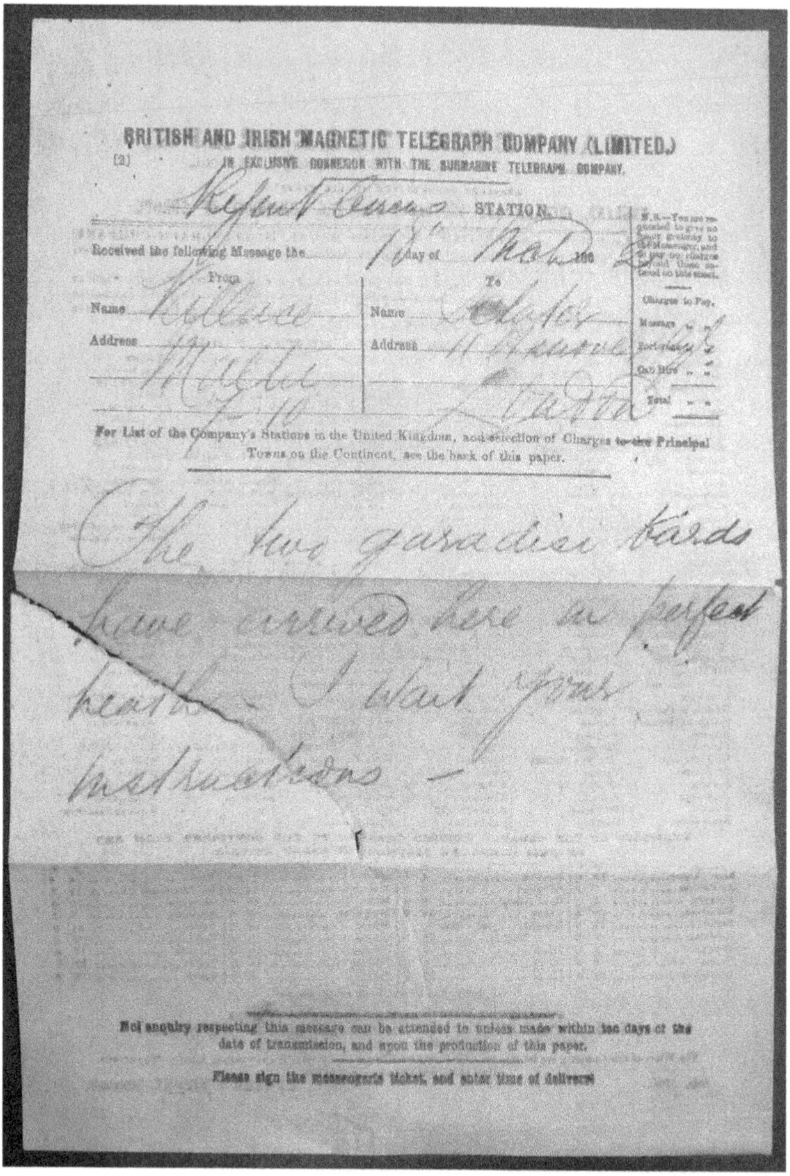

Fig. 25. Telegram of March 1862, sent from Malta

# Malta

## 87. To Philip Lutley Sclater, 18 March 1862

British and Irish Magnetic Telegraph Company (Limited)
in exclusive connexion with the submarine telegraph company.
Regent Circus Station
Received the following Message the 18th day of March 1862.

From:   Name: Wallace

         Address: Malta

To:    Name: Sclater

         Address: 11 Hannover Square, London

The two garadisi bards have arrived here in perfect health. I wait your instructions —

## 88. To Philip Lutley Sclater, 31 March 1862

**Pavilion Hotel, Folkestone**

*March 31st. 1862*

My dear D^r. Sclater

I have great pleasure in announcing to you the prosperous termination of my journey & the safe arrival in England (I suppose for the first time) of the *Birds of Paradise*.[384]

I did not get any reply to my telegram from Marseilles so after 26 hours waiting came on by express train, slept one night in Paris & on here today. I shall leave by the 9 o'clock train tomorrow arriving at London Bridge at *noon*, & shall expect to

---

[384] The arrival of the birds on 1 April was announced by Sclater to the meeting of the Zoological Society of London on 8 April 1862. Sclater, P.L. 1862. Report on the acquisition of two living Paradisea papuana brought by A.R. Wallace. *Proceedings of the Zoological Society of London* (April 8): 123. Named at first *Paradisea papuanus*, they are known to have been two males of the Lesser Bird of Paradise, *Paradisaea minor* (Shaw, 1809).

meet you or some one from you with a van to receive the Birds. The cages they are now in are about 3 feet by 2 ft 2 in. & 3 feet high—so I suppose the two would hardly go in a cab. I had them in much larger cages as far as Malta & had these made there to get better accommodation on board ship & on Railways. I assure you that during the seven weeks since I left Singapore I have had endless trouble & great anxiety with them. The stay of a week each at Bombay & Malta was of great use in obtaining supplies of food & good accommodation for them.

My principal difficulty has been first in getting regular supplies of soft muci-laginous fruits & living insects for them, without which I do not think they will long remain in health. Bananas they had till Suez & melons at Malta, but now nothing. *Cockroaches* they are excessively fond of & I managed to get them fairly regularly till leaving Bombay, but could never get enough to lay in a stock, & from Bombay to the Red Sea had to catch them of an evening on the beams of the ship, no other method succeeding. —

In the middle of the red sea it turned cold & no more were to be had till I reached Malta, & I was rather nervous having to give them hard boiled eggs instead, which they are very fond of but which I doubt agreeing with them for long. At Malta I got a stock of cockroaches which lasted me to Paris & I hope you will immediately take steps to get a supply. They will also eat dry bread but I do not like to give them much of it.

They make an immense deal of dirt & I have had trouble to get them well cleaned. The places where I could put them on board most of the ships was either too cold or too dark. Crossing the desert I had to travel with them myself all night in the luggage train. In the numerous changing of vessels & going on shore to hotels (sometimes in the middle of the night) I have had plenty of trouble, & travelling through France with them going as baggage, I have had no end of work to get permission to look after them, as the railway officials would insist upon applying to them the rules & regulations of ordinary baggage.

They have stood the cold wonderfully, having been in a temperature below 62° ever since we left Suez, corroborating my opinion expressed before leaving Singapore that cleanliness & fine air were of more necessity than a high temperature.

Fig. 26. The two male Lesser Birds of Paradise brought by Wallace to England exhibited at the Zoological Society of London

Their side plumes are about half-grown. When I left Singapore they were hardly visible having just moulted; they grew rapidly as far as Suez, the cold seemed to check them & I doubt if they will obtain their full development this year. Another year with a general temperature, *flying room, foliage*, & abundance of food, I hope they will be glorious.

I must now conclude

Remaining till tomorrow, my dear Sir

Yours very faithfully

Alfred R. Wallace

*P.L. Sclater* Esq.

# APPENDIX I

LETTER DETAILS[385]

## England to Singapore and Malacca

1. Wallace to George Silk, 19 and 26 March 1854. (Copy of letter 7 pp.) NHM WP1/3/27. WCP352. Printed: ML1: 332–3, Marchant 1: 45–7.

2. Wallace to Mary Ann Wallace, 30 April 1854. (ALS 4 pp.) NHM WP1/3/28. WCP353. Printed: ML1: 48.

3. Wallace to [Edward Newman], 9 May 1854. WCP4259. Printed: *Zoologist* 12 (142) (August 1854): 4395–7 [S14].

4. Wallace to Mary Ann Wallace, 28 May 1854. (ALS 3 pp.) NHM WP1/3/29. WCP354. Printed: ML1: 337–8, Marchant 1: 48.

5. Wallace to Mary Ann Wallace, [24] July 1854. (ALS 4 pp.) NHM WP1/3/30. WCP355. Printed: ML1: 338, Marchant 1: 49.

6. Wallace to Mary Ann Wallace, 30 September 1854. (ALS 4 pp.) NHM WP1/3/32. WCP357. Printed: ML1: 340, Marchant 1: 51.

7. Wallace to George Silk, 15 October [1854]. (ALS 2 pp.) NHM WP1/3/33. WCP358. Printed: Marchant 1: 52.

---

[385] Where relevant, the text in brackets following the date describes the item using standard archival abbreviations. Hence (ALS 4 pp.) indicates that this item is an autograph letter, signed by Wallace, with an extent of 4 pages.

## Borneo

8. Wallace to Henry Norton Shaw (RGS), 1 November 1854. (ALS 4 pp.) Royal Geographical Society, Letter book 1854 (JMS 8/17). WCP3554.

9. Wallace to Samuel Stevens, [c. December 1854]. Printed: *Proceedings of the Entomological Society of London,* in the *Transactions* (new series) 3: 87 [session of 2 April 1855].

10. Wallace to Samuel Stevens, 8 April 1855. WCP4261. Printed: *Zoologist* 13 (154) (August 1855): 4803–7 [S21].

11. Wallace to John Wallace, 20 April 1855. (Typed copy, 1 p.) California Historical Society. WCP1829.

12. Wallace to George Robert Waterhouse, 8 May 1855. (ALS 4 pp.) NHM-Catkey-418279. WCP781.

13. Wallace to Frances Sims, 25 June 1855. (ALS 6 pp.) NHM WP1/3/34. WCP359. Printed: ML1: 343–5, Marchant 1: 56.

14. Wallace to Frances Sims, 28 September 1855 and 17 October 1855. (ALS 4 pp.) NHM WP1/3/35. WCP360.

15. Wallace to Mary Ann Wallace, 25 December 1855. (ALS 4 pp.) NHM WP1/3/36. WCP361. Printed: ML1: 345.

16. Charles Darwin to Wallace, [December 1855]. (draft 1 p. and list 2 pp.) CUL DAR 206.34–35. WCP4758. Printed: CCD5: 512 (no. 1812).

17. Wallace to Frances Sims, 20 February [1856]. (ALS 4 pp.) NHM WP1/3/37. WCP362. Printed: ML1: 347, Marchant 1: 60–1.

18. Wallace to Thomas Sims, [c. 20 February 1856]. (ALS 4 pp.) NHM WP1/3/61. WCP385. Printed: ML1: 348, Marchant 1: 61.

19. Wallace to Samuel Stevens, 10 March 1856. (ALS 4 pp.) CUL Add. 7339/232. WCP1701.

20. Wallace to Frances Sims, 21 April 1856. (ALS 4 pp.) NHM WP1/3/38. WCP363. Printed: Marchant 1: 62.

21. Wallace to [Frances and Thomas Sims], [21 April 1856]. (ALS 2 pp.) NHM WP1/3/64. WCP388.

22. Wallace to Henry Walter Bates, 30 April and 10 May 1856. (ALS 7 pp.) NHM WP1/3/39 (original; also a copy in another handwriting, 15 pp.). WCP364. Printed: ML1:350.

23. Wallace to Samuel Stevens, 12 May 1856. (ALS 4 pp.) CUL Add. 7339/233. WCP1702.

## Bali, Lombock, and Celebes

24. James Brooke to Wallace, 4 July 1856. (ALS 7 pp.) BL Add. 46441 ff. 2–5. WCP3073.

25. Wallace to Samuel Stevens, 21 August 1856. (ALS 4 pp.) CUL Add. 7339/234. WCP1703. Printed: *Zoologist* 15 (171) (January 1857): 5414–16 [S31].

26. Wallace to Samuel Stevens, 27 September 1856. (ALS 4 pp.) CUL Add. 7339/235. WCP1704. Printed: *Zoologist* 15 (176) (May 1857): 5559–60 [S32].

27. James Brooke to Wallace, 5 November 1856. (ALS 7 pp.) BL Add. 46441 ff. 6–8. WCP3074.

28. Henry Walter Bates to Wallace, 19 and 23 November 1856. (Copy of letter 14 pp.) NHM Catkey-418383. WCP824. Printed: Marchant 1: 64–5.

29. Wallace to Samuel Stevens, 1 December 1856. WCP4262. Printed: *Zoologist* 15 (179) (July 1856): 5652–7 [S33].

30. Wallace to John and Mary Wallace, 6 December 1856. (Typed copy, 1 p.) California Historical Society. WCP1828.

31. Wallace to Frances Sims, 10 December 1856. (ALS 5 pp.) NHM WP1/3/40. WCP365. Printed: Marchant 1: 64.

## Aru and Amboyna

32. Wallace to Samuel Stevens, 10 March and 15 May 1857. WCP4746. Printed: *Proceedings of the Entomological Society of London* 1857 [meeting of 5 October 1857]: 91–3 [S35].

33. Charles Darwin to Wallace, 1 May 1857. (ALS 8 pp.) BL Add. 46434 ff. 1–4. WCP1839. Printed: ML2: 95, Marchant 1: 129, LLD2: 95, CCD6: 387–8 (no. 2086).

34. Wallace to Henry Norton Shaw, August 1857. (copy of letter 2 pp.) Royal Geographical Society, Corr. Block CB4 1851–60. WCP3561.

35. Wallace to Charles Darwin, [27 September 1857]. (AL 2 pp.) CUL DAR 47.145. WCP4080. Printed: CCD6: 457–8 (no. 2145).

36. James Brooke to Wallace, 31 October 1857. (ALS 4 pp.) NHM WP1/9/32. WCP1532. Printed: MA1: 98.

37. Wallace to Samuel Stevens, 20 December 1857. WCP4747. Printed: *Zoologist* 16 (191) (June-July 1858): 6120–4 [S44].

38. Charles Darwin to Wallace, 22 December 1857. (ALS 8 pp.) BL Add. 46434 ff. 5–8. WCP1840. Printed: Marchant 1: 131, LLD2: 108, CCD6: 514–15 (no. 2192).

39. Wallace to Philip Lutley Sclater, [January 1858]. (ALS 2 pp.) Zoological Society of London, enclosed with letter to P.L. Sclater of 4 April 1862. WCP1724.

40. Wallace to Henry Walter Bates, 4 and 25 January 1858. (ALS 4 pp., with enclosure 2 pp.) NHM WP1/3/41 (manuscript copy NHM-Catkey-418384) and NHM WP1/3/72. WCP366. Printed: ML1: 358, Marchant 1: 65.

## Ternate and New Guinea

41. Wallace to Frederick Bates, 2 March 1858. (ALS 4 pp.) NHM WP1/3/42 (NHM Catkey-418388: copy in another handwriting, 2 pp.). WCP367. Printed: Marchant 1: 69.

42. James Motley to Wallace, 22 May 1858. (ALS 4 pp.) BL Add. 46435 ff. 2–3. WCP2024.

43. Charles Darwin to Wallace, [13 July 1858]: Sketch of Mr. Darwin's 'Natural Selection'. (Note 1 p.) Wallace *Notebook 4*, p. [14b] in Linnean Society of London.

44. Wallace to Samuel Stevens, 2 September 1858. WCP4274. Printed: *Zoologist* 17 (200) (March 1859): 6409–13 [S45], partly in *Ibis* 1 (1) (January 1859): 111–13.

45. Wallace to Henry Norton Shaw, September 1858. (L 1 p.) WCP3562. Royal Geographical Society, Corr. Block CB4 1851–60.

46. Wallace to Frances and Thomas Sims, 6 September 1858. (ALS 4 pp.) NHM WP1/3/43. WCP368.

47. Wallace to [Samuel Stevens], 5 October 1858: 'Direction for Collecting in the Tropics'. (Note, 2 pp.) CUL DAR 270.1.2. WCP4068.

48. Wallace to Mary Ann Wallace, 6 October 1858. (Copy of letter (extracts), 4 pp.) NHM WP1/3/44. WCP369. Printed: ML1: 365, Marchant 1: 71–2.

49. Wallace to Joseph Dalton Hooker, 6 October 1858. (ALS 3 pp.) CUL Add. 7339/237, CUL DAR270.3.1 (Quentin Keynes bequest). WCP1454. Printed: CCD7: 166 (no. 2337).

## Batchian and Ternate

50. Wallace to Samuel Stevens, 29 October 1858. (ALS 4 pp.) CUL Add. 7339/236. WCP1705. Printed: *Proceedings of the Entomological Society of London* 1858–1859: 61; *Proceedings of the Zoological Society of London* 27 (22 March 1859): 129; *Ibis* 1 (2) (April 1859): 209–11; *Zoologist* 17 (204) (June 1859): 6546–7; *Annals & Magazine of Natural History* (3rd s.) 5 (26) (Feb. 1860): 145 [S47, S48].

51. Wallace to Francis Polkingthorne Pascoe, 28 November 1858. (ALS 4 pp.) NHM WP1/8/262. WCP1462.

52. Wallace to George Silk, 30 November 1858. (ALS 4 pp.) NHM WP1/3/45. WCP370. Printed: ML1: 365. Enclosure: 'Note on the smoke nuisance'. (Note, 3 pp.) NHM WP1/3/60. WCP370.

53. Charles Darwin to Wallace, 25 January 1859. (ALS 8 pp.) BL Add. 46434 ff. 9–12. WCP1841. Printed: Marchant 1: 145, LLD2: 134, CCD7: 240–1 (no. 2405).

54. Wallace to Samuel Stevens, 28 January 1859. WCP4749. Printed: *Proceedings of the Entomological Society of London* 1858–1859: 70; *Zoologist* 17 (206) (August 1859): 6621–2 [S50].

55. Charles Darwin to Wallace, 6 April 1859. (ALS 4 pp.) BL Add. 46434 ff. 13–14. WCP1842. Printed: ML1: 118, Marchant 1: 136, CCD7: 279 (no. 2449).

56. Wallace to Thomas Sims, 25 April 1859. (ALS 4 pp.) NHM WP1/3/46. WCP371. Printed: ML1: 367.

57. Daniel Hanbury to Wallace, 6 May 1859. (Copy of letter 1 p.) Wellcome Trust Library WMS 5304. WCP3527.

## Menado, Amboyna, and Ceram

58. Wallace to Francis Polkingthorne Pascoe, 20 July 1859. (ALS 4 pp.) NHM WP1/8/263. WCP1463.

59. Charles Darwin to Wallace, 9 August 1859. (ALS 6 pp.) BL Add. 46434 ff. 15–17. WCP1843. Printed: Marchant 1: 137, LLD2: 161, CCD7: 323–4 (no. 2480).

60. Wallace to John Gould, 30 September 1859. (ALS 2 pp.) NHM-Catkey-418781. WCP786. Printed: *Proceedings of the Zoological Society of London* 28 (24 January 1860): 61; *Annals and Magazine of Natural History* (3rd s.) (6) (July 1860): 75–6; *Zoologist* 19 (234) (September 1861): 7710 [S55].

61. Wallace to Philip Lutley Sclater, 22 October 1859. WCP4276. Printed: *Ibis* 2 (6) (April 1860): 197–9 [S58].

62. Charles Darwin to Wallace, 13 November 1859. (ALS 4 pp.) BL Add. 46434 ff. 18–19. WCP1844. Printed: Marchant 1: 139, LLD2: 220, CCD7: 375 (no. 2529).

63. Wallace to Henry Walter Bates, 25 November 1859. (ALS 4 pp.) NHM WP1/3/47. WCP372. Printed: ML1: 369, Marchant 1: 72.

64. Wallace to Samuel Stevens, 26 November 1859 and 31 December 1859. WCP4277, 4278. Printed: *Ibis* 2 (7) (July 1860): 305–6 [S59].

65. Wallace to Samuel Stevens, 14 February 1860. WCP4279. Printed: *Ibis* 2 (7) (July 1860): 306 [S59].

66. Daniel Hanbury to Wallace, 17 February 1860. (Copy of letter 1 p.) Wellcome Trust Library WMS 5304. WCP3528.

67. Charles Darwin to Wallace, 18 May 1860. (ALS 4 pp.) BL Add. 46434 ff. 21–22. WCP1846. Printed: Marchant 1: 141, LLD2: 309, CCD8: 219–21 (no. 2807).

## Waigiou, Ternate, and Timor

68. Wallace to George Silk, 1 September 1860 and 2 January 1861. (AL 8 pp., ALS 2 pp.) NHM WP1/3/48 and WP1/3/63. WCP373, 387. Printed: ML1: 371–3.

69. Wallace to Charles Darwin, [December 1860]. (ALS 2 pp.) CUL DAR 45.1b. WCP4079. Printed: CCD8: 504 (no. 2627).

70. Wallace to Samuel Stevens, 7 December [1860]. WCP4751. Printed: *Ibis* 3 (10) (April 1861): 211–12 [S61].

71. Wallace to Philip Lutley Sclater, 10 December 1860. WCP4762. Printed: *Ibis* 3 (11) (July 1861): 310–11 [S63].

72. Wallace to Francis Polkingthorne Pascoe, 20 December 1860. (ALS 4 pp.) Oxford Museum of Natural History (Hope Entomological Library), ARW238. WCP4317. Printed: *Proceedings of the Royal Entomological Society of London* 14 (1939): 77–8.

73. Wallace to Henry Walter Bates, 24 December 1860. (ALS 4 pp.) NHM WP1/3/49. WCP374. Printed: Marchant 1: 72, ML1: 373.

74. Wallace to Samuel Stevens, 6 February 1861. WCP4753. Printed: *Ibis* 3 (11) (July 1861): 311 [S63].

75. Wallace to Thomas Sims, 15 March 1861. (ALS 28 pp.) BL Add. 39168 ff. 2–27. WCP3351. Printed: Marchant 1: 73.

## Java, Sumatra, and back home

76. Wallace to Mary Ann Wallace, 20 July 1861. (ALS 3 pp.) NHM WP1/3/50. WCP375. Printed: Marchant 1: 83.

77. François de Caumont LaPorte to Wallace, 26 August 1861. NHM WP1/8/156. WCP1344.

78. Wallace to Philip Lutley Sclater, [11–15] and 20 September 1861. (ALS 4 pp.) Zoological Society of London, GB 0814 BADW. WCP1718. Printed: *Ibis* 4 (13) (January 1862): 95–6 [S66].

79. Wallace to Frances Sims, 10 October 1861. (ALS 4 pp.) NHM WP1/3/51. WCP376. Printed: Marchant 1: 85.

80. Wallace to Charles Darwin, 30 November 1861. (AL 4 pp.) CUL DAR 181.6. WCP4109. Printed: CCD9: 356–8 (no. 3334).

81. Wallace to Henry Walter Bates, 10 December 1861. (ALS 4 pp.) NHM WP1/3/52. WCP377. Printed: ML1: 377; Clodd, E. 1892. Memoir, in Bates, H.W. 1892. *The naturalist on the river Amazons*. London: John Murray, p. xxxiv.

82. Wallace to George Silk, 22 December 1861 and 20 January 1862. (ALS 10 pp.) NHM WP1/3/53. WCP378. Printed: ML1:379, Marchant 1: 87–8.

83. Mark Moss to Wallace, 3 February 1862. (ALS 1 p.) Zoological Society of London, GB 0814 BADW. WCP3684.

84. Mark Moss to Wallace, 6 February 1862. (Receipt 1 p.) Zoological Society of London, GB 0814 BADW. WCP3684.

85. Wallace to Philip Lutley Sclater, 7 February 1862. (ALS 4 pp.) Zoological Society of London, GB 0814 BADW. WCP1720.

86. Wallace to Samuel Stevens 16 February 1862. (ALS 4 pp.) Zoological Society of London, GB 0814 BADW. WCP1721.

87. Wallace to Philip Lutley Sclater, 18 March 1862. (Telegram 1 p.) Zoological Society of London, GB 0814 BADW. WCP1722.

88. Wallace to Philip Lutley Sclater, 31 March 1862. (ALS 3 pp.) Zoological Society of London, GB 0814 BADW. WCP1723.

# APPENDIX 2

## WALLACE'S ITINERARY IN THE MALAY ARCHIPELAGO

Wallace followed a quite complicated route through Southeast Asia, spending much time in Singapore, mainland Malaysia, and Borneo, followed by travels to the Moluccas and New Guinea, as well as the larger Indonesian islands of Sulawesi, Java, and Sumatra. His itinerary is here presented in summary form, where all localities within one island are combined into one entry. The dates are compiled from information in Wallace's publications, letters, and notebooks, as well as external data often regarding the movements of ships reported in newspapers. Many dates remain tentative, especially since some of Wallace's later recollections have proved inconsistent with other sources.

In this list, dates are given in the format: day, month (abbreviated), year. The length of stay is calculated according to the nights spent in the locality, hence the day of departure is not usually counted. Periods of travel between the islands have been omitted, as will be easily recognized from the sequence of the list. We have, however, provided details of his outward and homeward journeys on the steamers of the P&O company. The localities are listed according to the names used most frequently by Wallace in his writings, while the currently used designations and spellings are added in brackets to allow retrieval of the places on modern maps.

| | |
|---|---|
| 4 Mar 1854 | Departure from Southampton, on *Euxine* |
| 10 Mar 1854 | Gibraltar, on *Euxine* |
| 15 Mar 1854 | Malta, on *Euxine* |
| 20 Mar 1854 to 26 Mar 1854 [6 days] | Egypt: from Alexandria to Suez |
| 26 Mar 1854 | Departure from Suez on *Bengal* |
| 9 Apr 1854 to 10 Apr 1854 | Port de Galle (Sri Lanka) |
| 10 Apr 1854 | Departure from Port de Galle, on *Pottinger* |
| 17 Apr 1854 | Penang (Malaysia), on *Pottinger* |
| 18 Apr 1854 to 13 Jul 1854 [86 days] | Singapore (including one week on Pulau Ubin) |
| 15 Jul 1854 to 23 Sep 1854 [70 days] | Malacca (Malaysia) |
| 25 Sep 1854 to 17 Oct 1854 [22 days] | Singapore |
| 29 Oct 1854 to 10 Feb 1856 [539 days] | Sarawak (Borneo) |
| 17 Feb 1856 to 23 May 1856 [96 days] | Singapore |
| 13 Jun 1856 to 15 Jun 1856 [3 days] | Baly (Bali) |
| 17 Jun 1856 to 30 Aug 1856 [74 days] | Lombock (Lombok) |
| 2 Sep 1856 to 18 Dec 1856 [107 days] | Macassar (Makassar, South Sulawesi) |
| 31 Dec 1856 to 6 Jan 1857 [7 days] | Ke (Kai) Island |
| 8 Jan 1857 to 2 Jul 1857 [175 days] | Arru (Aru) Islands: Wamar, Wokam, Maikoor |

12 Jul 1857 to 15 Oct 1857    Macassar (Makassar, South Sulawesi)
[96 days]

15 Oct 1857 to 20 Oct 1857    Kaisa islands (Sulawesi)
[5 days]

20 Oct 1857 to 19 Nov 1857    Macassar (Makassar, South Sulawesi)
[30 days]

24 Nov 1857 to 26 Nov 1857    Timor (Kupang)
[2 days]

28 Nov 1857 to 29 Nov 1857    Banda Island (Banda Besar)
[2 days]

30 Nov 1857 to 4 Jan 1858    Amboyna (Ambon)
[35 days]

8 Jan 1858 to 19 Feb 1858    Ternate
[39 days]

19 Feb 1858 to 1 Mar 1858    Gilolo (Halmahera)
[10 days]

1 Mar 1858 to 25 Mar 1858    Ternate
[24 days]

26 Mar 1858    Makian (Machian) Island

28 Mar 1858 to 29 Mar 1858    Ganeh (Gani, South Halmahera)

11 Apr 1858    Mansinam, New Guinea

11 Apr 1858 to 29 Jul 1858    Dorey, New Guinea (Manokwari, Irian Jaya)
[109 days]

10 Aug 1858    Batchian (Bacan) Island

15 Aug 1858 to 13 Sep 1858    Ternate
[30 days]

14 Sep 1858 to 1 Oct 1858    Gilolo (Halmahera)
[18 days]

2 Oct 1858 to 9 Oct 1858    Ternate
[7 days]

| | |
|---|---|
| 9 Oct 1858 to 10 Oct 1858 [1 day] | Tidore Island |
| 10 Oct 1858 to 11 Oct 1858 [1 day] | March (Mare) Island |
| 11 Oct 1858 | Motir (Moti) Island |
| 11 Oct 1858 to 12 Oct 1858 [1 day] | Makian Island |
| 13 Oct 1858 to 20 Oct 1858 [7 days] | Kaioa Islands |
| 21 Oct 1858 to 21 Mar 1859 [151 days] | Batchian (Bacan) Island |
| 21 Mar 1859 to 1 Apr 1859 [11 days] | Kasserota (Kasiruta) Island |
| 1 Apr 1859 to 13 Apr 1859 [12 days] | Batchian (Bacan) Island |
| 20 Apr 1859 to 1 May 1859 [11 days] | Ternate |
| 5 May 1859 to 7 May 1859 [2 days] | Amboyna (Ambon) |
| 7 May 1859 to 11 May 1859 [4 days] | Banda Island (Banda Besar) |
| 13 May 1859 to 17 May 1859 [4 days] | Timor (Kupang) |
| 17 May 1859 to 21 May 1859 [4 days] | Semao (Semau) Island |
| 21 May 1859 to 27 May 1859 [6 days] | Timor (Kupang) |
| 31 May 1859 | Amboyna (Ambon) |
| 1 Jun 1859 | Banda Island (Banda Besar) |
| 7 Jun 1859 | Ternate |

| | |
|---|---|
| 10 Jun 1859 to 23 Sep 1859 [105 days] | Menado (NE Sulawesi) |
| 25 Sep 1859 to 27 Sep 1859 [2 days] | Ternate |
| 29 Sep 1859 to 29 Oct 1859 [30 days] | Amboyna (Ambon) |
| 31 Oct 1859 to 28 Dec 1859 [58 days] | Ceram (Seram) Island |
| 31 Dec 1859 to 24 Feb 1860 [55 days] | Amboyna (Ambon) |
| 26 Feb 1860 to 4 Apr 1860 [38 days] | Ceram (Seram) Island |
| 8 Apr 1860 to 12 Apr 1860 [4 days] | Manowolko (Manawoka), Gorong Islands |
| 12 Apr 1860 to 13 Apr 1860 [1 day] | Kissiwoi, Gorong Islands |
| 14 Apr 1860 to 16 Apr 1860 [2 days] | Uta, Gorong Islands |
| 17 Apr 1860 to 18 Apr 1860 [1 day] | Bam, Gorong Islands |
| 18 Apr 1860 to 25 Apr 1860 [7 days] | Matabello (Watoebela), Gorong Islands |
| 25 Apr 1860 | Manowolko (Manawoka), Gorong Islands |
| 26 Apr 1860 to 27 May 1860 [31 days] | Goram, Gorong Islands |
| 29 May 1860 to 1 Jun 1860 [3 days] | Kilwaru (Kiliwara) Island |
| 2 Jun 1860 to 17 Jun 1860 [14 days] | Ceram (Seram) Island |

23 Jun 1860 to 26 Jun 1860          Mesmon Islands (Jef Doif Islands)
[3 days]

4 Jul 1860 to 7 Aug 1860           Waigiou Island, New Guinea (Waigeo, Irian
[34 days]                                    Jaya)

7 Aug 1860 to 25 Sep 1860          Bessir, Gam Island (Irian Jaya)
[49 days]

25 Sep 1860 to 30 Sep 1860         Waigiou Island, New Guinea (Waigeo, Irian
[5 days]                                     Jaya)

1 Oct 1860 to 2 Oct 1860           Bessir, Gam Island (Irian Jaya)
[2 days]

3 Oct 1860 to 4 Oct 1860           Gagie Island, New Guinea (Gag, Irian Jaya)
[2 days]

11 Oct 1860 to 16 Oct 1860         Canidiluar, Gilolo (South Halmahera)
[5 days]

18 Oct 1860 to 21 Oct 1860         Ganeh, Gilolo (Gani, South Halmahera)
[3 days]

25 Oct 1860 to 1 Nov 1860          Kaioa Island
[7 days]

5 Nov 1860 to 2 Jan 1861           Ternate
[58 days]

7 Jan 1861 to 25 Apr 1861          Timor (Dili)
[108 days]

29 Apr 1861 to 1 May 1861          Banda Island (Banda Besar)
[2 days]

1 May 1861 to 3 May 1861           Amboyna (Ambon)
[3 days]

4 May 1861 to 2 Jul 1861           Bouru (Buru) Island
[59 days]

5 Jul 1861 to 6 Jul 1861           Ternate
[2 days]

| | |
|---|---|
| 7 Jul 1861 to 9 Jul 1861 [3 days] | Menado (NE Sulawesi) |
| 13 Jul 1861 to 14 Jul 1861 [1 day] | Macassar (Makassar, Sulawesi) |
| 16 Jul 1861 to 15 Sep 1861 [69 days] | Java (eastern part) |
| 18 Sep 1861 to 1 Nov 1861 [40 days] | Java (western part) |
| 2 Nov 1861 to 4 Nov 1861 [3 days] | Muntok (Bangka Island) |
| 8 Nov 1861 to 15 Jan 1862 [74 days] | Sumatra |
| 15 Jan 1862 to 16 Jan 1862 [1 day] | Muntok (Bangka Island) |
| 18 Jan 1862 to 8 Feb 1862 [21 days] | Singapore |
| 8 Feb 1862 | Departure from Singapore, on *Emeu* |
| 10 Feb 1862 | Penang (Malaysia), on *Emeu* |
| 16 Feb 1862 | Port de Galle (Sri Lanka), on *Emeu* |
| 22 Feb 1862 to 25 Feb 1862 [4 days] | Bombay (Mumbai, India) |
| 25 Feb 1862 | Departure from Bombay, on *Malta* |
| 12 Mar 1862 to 13 Mar 1862 [2 days] | Egypt: from Suez to Alexandria |
| 13 Mar 1862 | Departure from Alexandria, on *Ellora* |
| 17 Mar 1862 to 25 Mar 1862 [8 days] | Malta |
| 25 Mar 1862 to 30 Mar 1862 [6 days] | France: train from Marseilles to Boulogne |
| 31 Mar 1862 | Arrival in Folkestone |

# ITINERARY OF THE INDEPENDENT TRAVELS OF CHARLES ALLEN

| | |
|---|---|
| Jan 1860 to 24 Feb 1860 | Amboyna (Ambon) |
| 28 Feb 1860 to 28 Jun 1860 | Mysol (Misool) |
| 1 Jul 1860 to Sep 1860 | Ceram (Seram Island) |
| 15 Sep 1860 to 1 Oct 1860 | Mysol (Misool) |
| 15 Oct 1860 to 30 Oct 1860 | Ternate |
| 1 Nov 1860 to 15 Dec 1860 | Halmahera and Morotai Island |
| 16 Dec 1860 to 29 Dec 1860 | Ternate |
| 10 Jan 1861 to 31 May 1861 | New Guinea |
| 10 Jun 1861 to 21 Oct 1861 | Bouru (Buru) Island |
| Jul 1861 to Aug 1861 | Sula Islands |
| 1 Sep 1861 to 10 Sep 1861 | Ternate |
| 1 Oct 1861 to 15 Dec 1861 | Flores and Solor |
| Jan 1862 | Macassar (Makassar, South Sulawesi) |
| Jan 1862 to Feb 1862 | Coti (Koti, East Borneo) |
| 14 Feb 1862 | Arrival in Singapore |

Wallace employed Charles Allen as an independent collector from January 1860 to February 1862. As Allen apparently did not keep a daily log, all dates relating to his travels are uncertain and approximate. Dates are provided here merely to gain a general picture, but few can be verified exactly. It is likely that Wallace only met Allen on four occasions during this period: Amboyna in January 1860, Ternate in November 1860, and again in December 1860, and Bouru in June 1861.

# ABBREVIATIONS

AL          Autograph letter[386]

ALS         Autograph letter signed[387]

BL          British Library, London

CCD         Burkhardt, F.H. et al. eds. 1985–. *The correspondence of Charles Darwin*. 19 vols

CUL         Cambridge University Library

LLD         Darwin, F. ed. 1887. *The life and letters of Charles Darwin*. 3 vols

Lyell       Lyell, C. 1835. *Principles of geology*. 4 vols

Marchant    Marchant, J. ed. 1916. *Alfred Russel Wallace letters and reminiscences*. 2 vols

MA          Wallace, A.R. 1869. *The Malay Archipelago*. 2 vols

ML          Wallace, A.R. 1905. *My life*. 2 vols

NHM         Natural History Museum, London

RGS         Royal Geographical Society of London

S + no.     Smith, C.H. ed. 1998–. *The Alfred Russel Wallace page* (plus letter number)

WCP + no.   Wallace Correspondence Project (plus letter number)

---

[386] A handwritten, unsigned letter.
[387] A handwritten letter signed by the author.

# ILLUSTRATION CREDITS

1   Reproduced by kind permission of P&O Heritage Collection
2   A. R. Wallace Memorial Fund & G. W. Beccaloni
3   Natural History Museum/Science Photo Library
4a  Natural History Museum, London
4b  Natural History Museum, London
5   National Portrait Gallery, London
6   Royal Entomological Society
7   Image taken from *Chispa*, vol. 10, no. 1 (1970): 329. Reproduced with kind permission of Tuolumne County Historical Society
8   National Portrait Gallery, London
9   Science Photo Library
10  Natural History Museum, London
11  Tunbridge Wells Museum and Art Gallery
12  Natural History Museum/Science Photo Library
13  Natural History Museum, London
14  Royal Geographical Society
15  National Portrait Gallery, London
16  Smithsonian Institution Libraries
17  Natural History Museum, London
18  Natural History Museum/Science Photo Library
19  Royal Entomological Society
20  Natural History Museum, London
21  National Portrait Gallery, London
22  National Portrait Gallery, London

23  Natural History Museum, London

24  © National Portrait Gallery, London

25  Zoological Society of London

26  Illustrated London News, April 1862, Supplement p. 375.


Plate section   Heritage Editions

# BIBLIOGRAPHY

## MANUSCRIPTS

*Journal 1*: Linnean Society of London MS178a.
*Journal 2*: Linnean Society of London MS178b.
*Journal 3*: Linnean Society of London MS178c.
*Journal 4*: Linnean Society of London MS178d.
*Notebook 1*: Linnean Society of London MS179.
*Notebook 2/3*: Natural History Museum (London) Z MSS 89 O WAL.
*Notebook 4*: Linnean Society of London MS180.
*Notebook 5*: Natural History Museum (London) Z MSS 89 O WAL.

## REFERENCES

Baker, D.B. 2001. Alfred Russel Wallace's record of his consignments to Samuel Stevens, 1854–1861. *Zoologische Mededelingen* 75 (16): 251–341.

Beccaloni, G.W. ed. 2012–. *Wallace Letters Online* <http://www.nhm.ac.uk/wallacelettersonline>.

Beddall, B.G. 1988. Wallace's annotated copy of Darwin's *Origin of Species*. *Journal of the History of Biology* 21 (2): 265–89.

Burkhardt, F.H. et al. eds. [CCD] 1985–. *The correspondence of Charles Darwin*. Cambridge University Press. 19 vols.

Darwin, F. ed. [LLD] 1887. *The life and letters of Charles Darwin, including an autobiographical chapter.* 3 vols. London: John Murray.

Lyell, C. 1835. *Principles of geology: being an inquiry how far the former changes of the Earth's surface are referable to causes now in operation.* 4th ed. 4 vols. London: John Murray.

Marchant, J. ed. 1916. *Alfred Russel Wallace letters and reminiscences.* 2 vols. London: Cassell.

Rookmaaker, K. and Wyhe, J. van. 2012. In Alfred Russel Wallace's shadow: his forgotten assistant, Charles Allen (1839–1892). *Journal of the Malaysian Branch of the Royal Asiatic Society* 85 (2): 17–54.

Smith, C.H. and Beccaloni, G. eds. 2008. *Natural selection and beyond: the intellectual legacy of Alfred Russel Wallace.* Oxford University Press.

Smith, C.H. ed. [S] 1998–. *The Alfred Russel Wallace page and Bibliography of the Writings of Alfred Russel Wallace*: <http://people.wku.edu/charles.smith/index1.htm>

St. John, S. 1863. *Life in the forests of the far east.* London: Smith, Elder, & Co.

Wallace, A.R. [Sarawak Law] 1855. On the law which has regulated the introduction of new species. *Annals and Magazine of Natural History* (ser. 2) 16 (93, September): 184–96.

Wallace, A.R. [Ternate essay] 1858. On the tendency of species to form varieties; and on the perpetuation of varieties and species by natural means of selection. *Journal of the Proceedings of the Linnean Society of London. Zoology* 3 (20 August): 46–50.

Wallace, A.R. [MA] 1869. *The Malay Archipelago: the land of the Orang-utan, and the Bird of Paradise. A narrative of travel, with studies of man and nature.* 1st ed. 2. vols. London: Macmillan and Co.

Wallace, A.R. [ML] 1905. *My life: a record of events and opinions.* 2 vols. London: Chapman & Hall.

Wyhe, J. van and Rookmaaker, K. 2012. A new theory to explain the receipt of Wallace's Ternate essay by Darwin in 1858. *Biological Journal of the Linnean Society* 105 (1): 249–52.

Wyhe, J. van. ed. 2002–. *The complete work of Charles Darwin Online.* <http://darwin-online.org.uk/>.

Wyhe, J. van. ed. 2012–. *Wallace online*: <http://wallace-online.org/>

Wyhe, J. van. 2013. *Dispelling the darkness: voyage in the Malay Archipelago and the discovery of evolution by Wallace and Darwin.* Singapore: World Scientific Publishing.

Wyhe, John van & Kees Rookmaaker. 2013. Wallace's mystery flycatcher. *Raffles Bulletin of Zoology* 61 (1): 1–5, figs. 1–3.

Wyhe, John van. 2014. A delicate adjustment: Wallace and Bates on the Amazon and "the problem of the origin of species". *Journal of the History of Biology* vol. 47, issue 4: 627–659.

Wyhe, John van ed. 2015. *The Annotated Malay Archipelago by Alfred Russel Wallace.* Singapore: NUS Press.

# INDEX

Numbers in bold indicate page with illustration

Acarus [mite] 222
Adie's sympiesometer 33
Adolias 72
Agassiz, Louis 183, 225
Agelasta 112
Agia 75
Alcides 39
Alexandria xvi, 6–10, 256, 296, 301
Ali (Wallace's assistant) 31–2, 84–6, 121–2, 124, 183, 222, 230, 255
Allen, Charles Martin xv, 3, 11, 17, 18, 20, 23, 28, 31, 90, 101, 223, 227, 228, 229, 235, 248, 262–3
    itinerary 302
alligator 166
*Alma* schooner 86
Amazon valley 72, 96, 102–7, 125, 163–4, 192, 220, 239
Amberbaki 169, 170
*Ambon* steamer 254
Amboyna 121–51, 207–26, 297–300, 302
amok 97
Ampanam, Lombock 84, 91
Anthicus 106, 135, 136, 137, 145
Anthrax 137
Anthribidae [fungus weevils] 36, 42, 73, 137, 144, 238
*Antilla* brig 127
ants 16, 23, 38, 40, 171, 268
Aphodius 114

Apoderus 39
Aprosmictus amboinensis 237, 260
Arachnida 146
Areca, see Oreca palm
Argus pheasant 27
Aronias 162
Arru, see Aru Islands
Artamidae [woodswallows] 92
Artocarpus [jack fruit] 135, 145
Aru Islands 115–6, 117, 121–51, 296
Astathes 36, 112
Atrapia nigra 174
Ayer Panas 27

babirusa 260–1
Bacan Island, see Batchian
bacon 50, 54, 55
Baderoon (Wallace's assistant) 121, 122
Bali 79, 84–120, 296
Bally, see Bali
Baly, John Sugar 210
Bam Island 299
Banda Island 124, 127, 207, 297–8, 300
*Bangalore* ship 65
barbet 111
Basilornis 223
Baso (Wallace's assistant) 121, 122
Batavia xvii, 5, 34, 142, 167, 187, 244, 254, 259, 262
Batchian Island 179, 183–206, 298

Bates, Frederick xxi, 153, 157, **161**, 220, 272
Bates, Henry Walter xxi, 65, 71, **72**, 83, 86–7, 102, 123, 124, 139, 143, 151, 220, 228, 239, 269
Batocera 74
bee-eater 94
bees 16, 38, 102, 114, 198
Belionota sumptuosa 37, 39
Bembidiidae 159
*Bengal* steamer 4, 6, 7, 10, 296
Bentham, Jeremy 193, 266
Bernstein, Heinrich Agathon 236
Bessir 227, 230, 233, 300
Bileling, Bali 84
bird nests 198
bird of paradise, greater 121–2
    red 228
birds of paradise
    Aru 119, 121–51, 169–170
    Batchian 184–5, 187, 199
    Waigiou 227
    for sale in Singapore 255, 278–80
    taken alive to England 278–87, **282, 285**
Birmania 259
Blisset, John, gunmaker 98
Blondin 275
Blyth, Edward 43, 82, 139
Boisduval, Jean Baptiste A.D. de 15, 112, 114, 192, 273
Bombay xvi, 7, 256, 280–1, 301
Bonaparte, Charles Lucien 80, 216, 217, 261
Bontyne 96, 115, 120
Borneo 28–83
Borneo Company Ltd. 90, 100
Bostrichidae [auger beetles] 172
bottle used for insect collecting 38
Bouru 229, 249, 260, 300, 302
Bowerbank, James Scott 94
Bowring, John Charles 80, 145, 163, 171
Brachelytra 76, 135, 136, 171
Brachinidae 105, 145
Bree, Charles Robert 268
Brenthidae [straight-snouted weevils] 37, 144, 240

British Association for the Advancement of Science 250
British Museum 244–5
British Museum catalogues 64, 79
British Ornithologists Union 260
Bronn, Heinrich Georg 224
Brooke, Capt. John 26, 49, 98
Brooke, Charles Anthoni Johnson 133
Brooke, James xxi, 5, 23, 26, 28, 34, 43, 49, 55–6, 86, 88, **89**, 99, 130, 133
Bruné, see Brunei
Brunei 29, 90, 100
Buceros cassidix 111
Buceros lunatus 260, 263
Buch, C.L. von 228, 233
Bukit Timah 4–5, 11
Bulimus 111, 112
Buprestidae [wood-boring beetles] 16, 37, 38, 39, 40, 41, 75, 77, 78, 103, 137, 144, 146, 149, 160, 172, 177, 200, 240
Buprestis 40, 126, 221
Burmeister, Karl Hermann Konrad 146, 162
butterflies, see Lepidoptera
butterfly sent home 23

Cacatua citrinocristata 260–1
Calamus [rattan palm] 77
Calleamura 72
Callichroma 40
Callitheas 72
Caloenas nicobarica 216
Calyptomena 111
Cambodia 21, 23, 189, 258–9
Cambridge Philosophical Society 225
Canidiluar 300
Carabidae [ground beetles] 16, 37, 75, 77, 104, 112, 114, 135, 136, 137, 144, 145, 159, 163, 171, 177, 221
Carabus 146, 163
Carpophaga perspicillata 136, 218
Carter, Joseph 93
Casnonia 112, 159
cassowary 126, 217, 261–2

Casteleyns, J. 259
Castelnau, Francis de xxii, 220, 259
Catadromus 158
Catadromus tenebrioides 114
Catagramma 72, 109
Catascopus 37, 75, 159, 171
Catoxantha bicolor 75
Celebes crested macaque 187
centipedes 16
Ceram 189, 207–26, 299, 302
Cerambycidae 74, 102, 144, 160
Cerastonema wallacei 42, 73, **74**
Cethosia 14
Cetonia 40, 126, 137, 144, 170, 178
Cetoniadae [rose chafers] 37, 75, 77,
     105, 113
Chalcopsitta 263
Chalicodoma pluto 184
Chambers, Walter 90
Charaxes 72
Cheirotomus 83
China 6, 69
Chrysobothris 37
Chrysodema lotinii 172
Chrysomela 91, 136
Chrysomelidae [leaf beetles] 16, 37, 135, 172
Cicero 274
Cicindela 75, 91, 112, 126, 144, 158, 160,
     162, 170, 221, 234
Cicindela d'urvillei 170
Cicindela elegans 158
Cicindela funerata 170, 187
Cicindela heros 112, 158, 161
Cicindela tenuipes 160
Cicindelidae [tiger beetles] 16, 75, 104,
     160, 163, 210
Cinnyris 96
City of Bristol ship 94
Clarendon, Earl of 34, 60, 88
Clark, William 225
Claudet, Antoine François Jean 53, 176,
     204, 250
Cleridae [chequered beetles] 16, 37, 75,
     109, 149, 172, 189, 240

Clerus 144
Clytus 36, 40, 74, 105
Coccinella [ladybirds] 91, 102, 106
Cochinchina 69
cockatoo 82, 92, 128, 189, 237
Cocytia d'Urvillei 125
Coleoptera [beetles] 14, 16, 18, 38, 73, 76,
     91, 104, 112, 135, 144, 164, 220, 222
collections 26, 73–4, 203, 272
     birds 67, 77, 83, 94, 111, 223
     insects 67, 77, 94, 111, 150
     mammals 67, 94
     plants 57
     shells 94, 111
     sponges 94
collections
     Amboyna 146, 150, 160, 190, 192, 223
     Aru 143–6, 150, 160, 180, 190
     Borneo 59
     Ceram 223, 238
     Gilolo 216–18
     Java 265
     Lombock 94
     Macassar 111, 134, 150
     Malacca 150
     Mysol 238
     New Guinea 168–74
     Sarawak 65, 81, 150
     Singapore 150
     Ternate 216–18
     Waigiou 232, 234–5, 238
     Sumatra 272
Colliuris 36, 37, 40, 112, 126
Collyris 75, 158, 159, 238
colonisation of islands 213
Coquille voyage 128
coracora 207
coral reefs 140
Cornubia ship 64
Coryphocera 38
Coti 302
Coulson, Robert 29, 36, 100
Coupang (Timor) xvii, 124
Crawford's Malay Dictionary 64

Crimean war 59, 63
Crookshank, Arthur Chichester 99
crown pigeon 174
Crystal Palace exhibition 22, 142, 193,
    247, 272, 274
Curculio 135, 136, 177, 221
Curculionidae [weevils] 16, 37, 39, 41, 42,
    102, 106, 126, 145, 170
Cybdelis 72, 107
Cyclica 102, 145, 149
Cyclostoma 111, 112
Cynopithecus nigrescens 187

dammar 223
Danaidae [milkweed butterflies] 96, 150
Danais 14, 15, 16
Darwin, Charles xxii, 30–1, **58**, 86, 87,
    94, 123, 129, 132, 139, 147, 156, 167,
    176, 180, 181, 185–6, 197, 201, 211,
    218, 224, 266
    notebook on evolution 129
    *Origin of Species* 197, 201, 208–9, 218,
    224, 228, 229, 231, 233, 239, 245–7,
    255, 266
    Sketch of Natural Selection 167–8
Darwin, Erasmus Alvey 226
Davis, Joseph Barnard 68, 193, 198
Dawson, John William 225
Deal, J. 178
Dejean, P.F.M.A. 162
Delli (Timor) xvii, 124, 241, 249, 266
Demetrias 75
Dickens, Charles 49, 63, 101, 275
Dillwyn, Lewis Llewelyn xxiii, 165–6
Diopsis 16
Diptera (flies) 16, 82, 103, 109, 114, 135,
    137, 145, 146, 150, 172, 221
Diurnes [true butterflies] 15, 107, 109, 150
Dobbo 121–2, 125, 127
Doleschall, Carl Ludwig 142, 143, 146
Dorey 155, 169–73, 187, 190, 191, 217, 218,
    223, 234, 259, 262, 297
Dracaena 81
dragon flies 16

Draper, Miss 52
Dromius 75
Drusilla 126
Dufour, P. 231
Duivenbode, Maarten Dirk van 151,
    167, 266
Dumas, A. 175, 193–4
*Dunedin* ship 80
durian 78
Dyaks 28, 44, 50, 55–6, 60, 165

Earl, Clara 90
Earl, George Windsor 90, 212
East India Railway 247
Ega, Amazon 102
Egypt 3, 6–10, 296
Elater 40, 75, 172, 234
Elateridae [click beetles] 16, 37, 103,
    145, 149
elephant 4, 22, 213, 275
*Eliza Thornton* barque 11, 18
*Ellora* steamer 301
Emesis 72
*Emeu* steamer 256, 301
Endomychidae 76
Entomological Society of London 35, 91
Eos cyanostriata 237, 262
Eos reticulata 237, 262
Eos squamata 237
Epicalias 72
Ergeteles 270
Erichson, Wilhelm Ferdinand 146
Erotylidae [fungus beetles] 16, 37, 76,
    102, 109, 144
Erycinidae [metalmark butterflies] 15, 72,
    102, 106, 128
*Etna* steamer 170
Euchirus [long-armed chafer beetle]
    98, 138
Euchirus longimanus 137, 143, 146
Euchlora 113–4
Euchroma gigantea 37
Eupholus cuvieri 170
Eupholus schoenherri 170

Euploea 14, 15, 16, 72, 115, 233
Eurycephalus maxillosus 40
Eurylaimus [broadbills] 111
*Euxine* paddle steamer xv, 3, **4**, 5, 10, 296
evolution 88, 122, 124–5, 161, 181

Family Herald 176, 193, 203, 250
Farnham 131
Favre, Pierre Etienne Lazare 20
Fechter, Charles Albert 275
Felis of Timor 213
Fenton, Roger 59
Fernandez, Manuel 84, 86
ferns 22, 66, 67, 264
fish 80, 82, 97, 119, 271
flies 91
Flores 302
flycatchers 16, 111, 261
Folkestone, Pavillion Hotel 283, 301
Forbes, Edward 140, 147, 213
Forsten, Eltis Alegondas 262
Fortnum & Mason 54
Foxcroft, James 177, 188
Fry, Herbert 177
Fry, Samuel 247

Gading 4, 19
Gagie Island 300
Galerucidae 172
gambir 11, 14
Ganeh 297, 300
Garo (assistant of Wallace) 183
Gärtner, Karl Friedrich von 130
Geach, Frederick F. 228, **248**, 255
Geodephaga [carnivorous
    ground-beetles] 75, 112, 126, 157,
    160, 163, 210
geographical distribution 29, 87, 92, 103,
    125, 147, 209, 211, 213, 238, 240, 246
Geological Society of London 132
Gibraltar 6, 296
Gilolo 128, 148, 151–5, 183, 216, 297
Glareola 261
Glenea 36, 39, **74**, 75, 210, 211

Gorilla war 275
Goram Island 299
Gorong Island 299
Gory, H.L. 163
Gosse, Philip Henry 164
Gould, Augustus Addison 141
Gould, John xxii, 80, 200, 208, 214, **215**,
    222, 235
Goura coronata 263
Goura victoriae 174
Grant, Charles Thomas Constantine 99
Gravelet-Blondin, Jean François 275
Gray, Asa 219, 245
Gray, George R. 162, 189, 200, 208, 228
Gray, John Edward 173, 179, 270
Gunung Ledang, see Mount Ophir
Gusti Ngurah Kketut Karang Asem 92
Guyana 269–71

Haggar, Morris 231
Hameraticherus, 75
Hamilton, Gray & Co. 12, 14, 80, 81, 83,
    98, 208, 249
Hanbury, Daniel xxii, 205, **206**, 223
Harpalidae 145
Harris, Mrs. of Dickens 49
Hart, Alfred Edward 228, 248, 266
Harvey Brand & Co. 97–8
Haughton, Samuel 225
Hayward, Mr. 7
Heekeren, Jacobus Johannes van 166
Heliconia 233
Heliconidae [heliconian butterflies] 15, 72
Helix 111
Helix glutinosa 112
Helms, Ludvig Verner 100
Hemiptera [bugs] 16, 91, 114
Heptadontas 158
Herbert, William 130
Hesperidae [skippers] 73
Hesthesis 40
Hestia d'Urvillei 125
Heteromera 37, 76
Hewitson, William Chapman 110, 273

Hill, Abraham 252
Hispidae 210
Hister 135
Home News 70
Hooker, Joseph Dalton 140, 156, **181**, 197, 199, 202, 213, 219, 225, 245
Hooker, William Jackson 26
hornbill 111
Horner, Ludwig 216
Horsfield, thomas 171
Huc, Evariste Régis 69
Huguenin, Otto F.U.J. 179, 183
Hunt, William 243
Huxley, Thomas Henry 125, 202, 245, 268
Hyades 126
hybridism 141
Hydrocissa exarata 111
Hydrophili 136
Hymenoptera [wasps and bees] 38, 114, 135, 137, 145

Ianthoenas halmaheira 217
Idaea 15
Ilkley Wells 219
Ips 187
Ixodinae 111

jack fruit 134, 135
jaguar 130, 132
Jardine, William 225
Jarvie, John 11, 12
Java 254–65, 301
Jobie island 175
Johnson, John Brooke 73, 99
Johnston, Alexander Laurie 98
Jukes, Joseph Beete 224
jungle cock 94

Kai, see Ké island
Kaioa Islands 183, 186, 298, 300
Kaisa Island 184, 217, 218, 296
Kalangan Banjermassing xxiii, 165
kangaroo 119, 173
Ké Island 121, 126, 296

Kasserota Island 298
Kasiruta Island 298
Kembang Djepoon bark 79, 84
Keyserling, Alexander von 224
Kilwaru Island 299
kingfisher, blue & white red billed 94 racquet-tailed 128
Kippist, Richard 188
Kissiwoi 299
Kölreuter, Joseph Gottlieb 130
Kraal, William 12, 81
Kuching 28

Labuan xxiii, 36, 59, 81, 90, 165, 259
Labuan Tring 91
Lacordaire, Théodore 65, 162, 163, 211, 220
Lahagi (assistant of Wallace) 183
Lahi (assistant of Wallace) 183
Lamellicornes [beetles] 14, 75, 102, 113, 126, 144, 149, 170, 238
Lamia 38, 102, 138, 144
Lamia hercules 113
Lamia octomaculata 113
Lamiadae 74
Lampyridae [fireflies] 172
Laphria 137
LaPorte, François de Caumont xxii, 258
Latchi (boatsman) 183
Latham, Robert G. 27
Lau Keng Tong 183
laurineous bark 205
law of priority 236
Lebra 75
Leicester, Collegiate College 252
leopard 130, 132
Lepidoptera [moths and butterflies] 15, 34, 72, 76, 82, 91, 114, 145, 177, 220, 222
Leptocercus 73
Leptocercus curius 16
Lepturidae 74
Lesson, René Primevère 128, 169, 262
Levaillant, François 173
Limentis 72

limestone hills 123, 134
Linnean Society of London 80, 152, 156, 185, 193, 207, 209, 221
Lobb, Thomas 81, 83, 101
Lobo Raman 255, 269, 273
locality tickets 77–8
Logan, James Richardson 15
Lomaptera 126, 170
Lombock 79, 84–120, 296
Longicornes [long-horned beetles] 16, 36–40, 74–8, 102, 105–10, 112, 126, 137, 144–6, 149, 160, 163, 170, 177, 186, 190, 200, 210, 221, 234, 237–8, 240, 272
Longicornia Malayana 160, 211
lories 82, 170, 189, 217, 222, 237
Lorius 237, 260
Lorius domicella 223, 237, 262
Lorius garrulus 217
Loudon, gunmaker 98–9
Lubbock, John 202
Lucanidae [stag beetles] 75, 138, 144
Lucanus 114, 145, 170, 178
Lycaenidae [gossamer-winged butterflies] 15, 73, 115, 126, 150, 170, 272
Lyell, Charles 30, 87, 139, 140, 154–6, 180, 181, 185, 188, 193, 197, 198, 202, 219, 224–6, 245–6

Macassar, Celebes 32, 60, 68, 71, 78, 79, 82, 84–120, 124, 234, 296–7, 301, 302
Macrocephalon maleo 207, 215
Macronota diardi 38, 40
Macropygia reinwardtii 216
Madeira 140, 213
Maikoor 296
Makian Island 183, 297, 298
Malacca 4, 19–22, 32, 72, 240
Malacodermes [soft-winged beetle] 76, 102, 149, 234
Malacomacrus 38, 76
maleo 207
Malta xvi, 6, 14, 256, 280–3, 296, 301

*Malta* steamer 301
Malthus, T. 154, 168, 201
Mamajam 95, 122
Mandeville, James W. 196
mangosteen 78
Manowolko 299
Mansinam 155, 297
Manyol, see Minszech
Mare Island 298
Maros, Celebes 98, 123, 134
marriage 41, 194, 277
Marseilles 222, 256, 281, 283, 301
mastodon 89, 213, 267
Matabello Island 299
Matthew, Patrick 226
Matthews, Sarah 53
Matthews, William 12
Mauduit, Anatole 4, 11, 17, 68
Maull, Henry 177
Mayall, Jabez Edwin Paisley 53
McDougall, Francis Thomas 59, 90
Mckinney, H. Lewis 151
Mecocerus gazella 37
Mecopus 39
Megacephala 75, 104
Megacephalon 215
Megachile pluto 184
Megalaima australis 263
Megalomma elegans 158
Megapodius 92, 215, 216, 237
Melandrya 76
Melastonia 14, 265
Melolonthidae [scarab beetles] 102, 114, 136, 238
Menado 113, 207–26, 299, 301
Mesman, Willem Leendert 86, 95, 98, 121
Mesmon Islands 300
mias, see orang utan
Micocerus gazella 42
Mimeta 229, 261
mimicry 228, 229
Minszech, Count George Vandalia 65
model republic 196–7
Moera Dua 254

Mohnike, Otto 124, 138, 146
molluscs 131, 141, 267
monkey of Batchian 187
    of Malacca 21
    of Sumatra 275–6
Monohammus 74, 146
Mormolyce 75
Moss, Mark xxii, 255, 278, 279
Motir Island 183, 298
Motley, James xxiii, 165–7
Mount Ophir 4, 22, 27, 33, 158
Muka 227
Müller, Salomon 216, 262
Mulsant, M.E. 163
Muntok 254, 301
Murchison, Roderick Impey 131
Murray, Andrew 225
Murray, Charles A. 30, 31
Murray, John 201, 218
Mysol 223, 227, 234, 237, 238, 261, 302

Natterer, Johann 220
natural selection 124, 151–3, 167, 198, 208,
    226, 230, 246
Natuurkundige Commissie voor
    Nederlandsch-Indië 216, 262
Necrophaga 136, 137, 145, 149
Nederlandsch Indische Commissie 170
Nederlandsche Handel Maatschappij 167
New Guinea 117, 151–82, 302
Newman, Edward 14, 66, 93, 97
Nicol, George Garden 11, 12
night collecting 172–3
Nitidula 102, 135, 145
Norton Shaw, Henry xxiii, 32, **33**, 131, 174
nutmeg 14
Nymphalidae [brush-footed
    butterflies] 72, 102, 103, 115, 135, 150

Odontocheila 104, 158
Onthophagus 135, 136, 145
orang utan 30, 36, 43, 46, 47–9, 66, 67,
    68, 77, 81–2, 97, 133, 166
Oreca palm 14

orioles 94, 229
Ornithoptera 73, 115, 125, 135, 145, 170,
    187, 199–200, 273
Ornithoptera amphimedon 135
Ornithoptera brookiana 73
Ornithoptera codrus 187
Ornithoptera croesus 184
Ornithoptera d'Urvilliana 127
Ornithoptera haliphron 135, 145
Ornithoptera helena 135
Ornithoptera poseidon 125
Ornithoptera remus 114, 135
Orthogonius 75, 158
Orthoptera 103, 150, 172, 221
Osculati, Gaetano 220
Otops 39
Owen, Richard 67, 97, 179, 213, 224,
    244, 268

Pachyrhynchus 42, 149, 187
Padday, Reginald 12, 100
Palembang 254, 272
Pall Mall 197, 274
palm trees 9, 13, 18, 77, 178
Pandani 77
Papilio 16, 72, 78, 96, 102, 105, 115, 125,
    150, 188, 210, 221, 233, 240, 269,
    271, 273
Papilio agamemnon 39
Papilio androcles 145
Papilio ascalaphus 115
Papilio codrus 80
Papilio deiphonus 137
Papilio dolicaon 273
Papilio encelades 134
Papilio eurypilus 115, 134
Papilio evemon 39, 40
Papilio helenus 115
Papilio iswara 40
Papilio peranthus [blue swallowtail] 91, 134
Papilio polyphontes 115
Papilio protesilaus 16, 273
Papilio rhesus 134
Papilio sarpedon 40, 115, 134, 187

Papilio severus 137
Papilio telemachus 187, 200
Papilio ulysses [blue mountain
    swallowtail] 126, 137, 146, 187
Papuan race 121, 193
Paradisea 199, 200
Paradisea apoda 127, 173, 214
Paradisea atra 236
Paradisea minor 283
Paradisea papuana 142, 169, 173, 280, 283
Paradisea regia 128, 169, 214
Paradisea rubra 169, 235
Paradisea superba 169, 236
Paradisea wallacii 185, 189
parrot 130, 217, 236
Pascoe, Francis Polkingthorne xxiii, 160,
    190, **191**, 209, 237
Passo 222
Pastor corythaix 111
Paterson Simons & Co 248
Paussus [ground beetles] 38, 76, 173
Payen, Antoine 65–6, 116
Penang xvi, 69, 256, 296, 301
Peninjau on Serambu Hill 28
Pericallus 36, 75
Pfeiffer, Ida Laura xiv, 49, 78, 80, 96
Phaenicophaus callirhynchus 111
Philhydrida 145
Phillips, John 225, 268
Phryneta 74
Phyllornes [bulbuls] 111
phytophaga 76, 114
Pica albicollis 111
Pictet, François Jules 224
Pieridae [white and sulphur
    butterflies] 91, 96, 102, 115, 150, 170,
    188, 221, 240
Pieris zaranda 134
pigeons 58, 86, 94, 96, 128, 130, 167, 217,
    222, 247, 261, 268
pins 22, 47, 166, 188
Pitta 94, 261
Pitta maxima 215–6
plantains 78, 264

Platycercus 217, 223
Plyctolophus sulphureus 92
Pompilidae 137
Port de Galle xvi, 256, 260, 296, 301
potato 120, 249, 276
*Pottinger* steamer 4, 10, 296
poultry 58, 86, 87, 94, 130
Powell, Baden 100
Pre-Raphaelite Brotherhood 243
Priamus 125, 146, 199, 200
Prionidae 74, 75, 106, 144
Prioniturus platurus 96
Prionus 138
Protaetia 113
Pselaphidae 106, 136, 145, 159
Psittacidae 217, 237, 262
Ptilonopus 216, 217, 261
Ptilotis 92
Pulau Ubin 32, 296
Punch (magazine) 27, 49

Quaritch, Bernard 80

Raja of Lombock 92
Ramsay, Andrew Crombie 224
Rappa, George 81, 255
raptorial birds 96, 111
rattan palm 77
Rejlander, Oscar Gustave 176
reptiles 66–7
resin 223
revolver 27
rhinoceros 4, 22, 27, 275
Rhyncophora [weevils] 42, 73, 114, 136,
    144, 149
rice 19, 22, 27, 84, 93, 119, 122, 130, 276
Richthofen, Ferdinand von 255, 267
Roberts, Eliza 120
Rogers, Henry Darwin 224
Rosenberg, Carl B.H. von 174, 262
Royal Geographical Society 32, 60
rug net 38
runner duck 86, 94
Rupicola 111

Sadong River 29, 30, 35, 44
Sala, George Henry Augustus 63
*Santubang* bark 31
Sarawak 26, 28–83, 88, 296
Sarawak law paper 29, 86, 89, 108–9, 123,
    129, 132, 139, 146, 155
Satyridae [brush-footed butterflies]
    14, 72, 102, 108, 150
Saunders, William Wilson 67, 77, 81, 93,
    159, 189, 198, 210, 221, 247
Scarabaeus atlas 113, 114
Scaritidae 105
Schaaffhausen, Hermann Joseph 226
Schneider, Carl F. A. 267
Sclater, Philip Lutley xxiii, 80, **142**, 216,
    219, 235, 255–6, 259, 279, 283
Scolytidae 102, 172
scorpions 16
Scythrops Novae-Hollandiae 111
Sebastopol 26, 60
Sedgwick, Adam 225
Semao Island 298
Semioptera wallacei 185, 214, 235
Serixia 187
shells 10, 26, 34, 59, 66, 67, 81, 94, 111,
    128, 140, 172, 180, 232
Si Munjon 29, 30, 35, 42, 51
Siam 189, 258
Siamang 281
Silk, George x, xxiv, 5, **6**, 11, 17, 22, 26,
    52, 57, 60, 118, 156, 175, 185, 192, 227,
    230, 255, 273
Simia morio 43
Sims, Edward 45–6, 120
Sims, Frances (Fanny) xxiv, **13**, 44, 51, 58,
    **61**, 62, 69, 70, 81, 118, 175, 257, 263
Sims, Thomas xxiv, 13, 17, 22, 45, 50, 51,
    59, 62, **63**, 70, 120, 175, 179, 203,
    229, 241
  in Albany Street 51, 70
  in Conduit Street 51–2, 70
Singapore 4, 10–19, 21, 26, 32, 58, 63, 64,
    67, 69, 71, 72, 98–9, 254–6, 278–87,
    296, 301
  Bukit Timah 4, 11, 14

London Hotel 11
  Hotel d'Europe 259
Smith, Frederick 198
smoke nuisance 185, 194, 195–6
Society for the Propagation of the Gospel
    in Foreign Lands 274
Society of Arts 185
Solor 302
Southampton xv, 281, 296
Sphaeridii 145
spider 195
Spix, Johann Baptist Ritter von 220
Spruce, Richard 26, 78, 143, 188
Spurgeons, Charles Haddon 176
Squire, Henry 179
St John, Spenser 28, 90, 100
Staphylinidae [rove beetles] 102, 105, 135,
    137, 145, 149, 171, 177
stereomonoscope 176
Stevens, Samuel xxiv, 12, 17, 22, 23, **35**,
    46, 49, 64, 71, 79, 86, 91, 95, 110,
    118, 125, 131, 134, 157, 168, 175, 177,
    186, 190, 191, 199, 220, 222, 223, 228,
    234, 241, 247, 257, 265, 280
stores needed in Aru 119
Strange, Frederick 50
subsidence 140
Suez xvi, 10, 256, 284, 296
Sula 263, 302
Sumatra 254–78
Surabaya 254, 256, 259
  Hotel der Nederlanden 258
Switzerland 192

Taeniodera 37, 113
Tak, Van der 167
Tanygnathus 260
Tanysiptera 217, 223, 261
tapir 275
Tatum, Thomas 210
Temminck, Coenraad Jacob 43, 82, 237
Temnosternus 112
Tenthredinidæ [sawflies] 16
Terias 14
Termos clarissa 72

Ternate 144, 148, 151–82, 183–206, 216, 227–53, 297–300, 302
Ternate essay xi, xv, 124, 153–6, 185, 188, 193, 198, 202, 208–9, 218, 255
Thecla 72, 108
Therates 37, 75, 112, 126, 158, 160, 170–1, 187, 234
Therates basalis 170
Therates fasciata 112
Therates festiva 171
Therates flavilabris 112, 158
Therates labiata 160
Thompson, J. 190
Thyreopterae [ground beetles] 37, 39, 75, 104,
Tidore Island 183, 298
tiger 4, 16–7, 22, 275
tiger beetle 31, 84, 124
Timalia 111
Timor 213, 227–53, 297–8, 300
Tmesisternus 200, 210, 238
Tmesisternus mirabilis 126
Tmesisternus septempunctatus 113
Todiramphus funebris 216
Tonquin 69
Triammatus 186
Trichius 113
Trichoglossus 237, 260
Tricondyla 75, 126, 158, 159
Tricondyla aptera 160, 171
Trimera 37, 149
trogon 111
Tropidorhynchus 92, 229, 261
trushes 111
Tunantins, Amazon 102, 107, 143, 271
Turner, Joseph M.W. 243

Uta Island 299

varieties 188
Vespidae 137
Vignoles, Charles Blacker 52

Waasbergen, Abraham van 116, 121–2
Waigiou 227–53, 300

Walford, Edward 177
Wallace Line 86–8, 125
Wallace, Alfred Russel
    itinerary 295–301
    photo 13, 248
    religious beliefs 251–52
Wallace, Frances (Fanny) see Sims, Frances
Wallace, John xxv, 18, 41, 50, 57, 117, 180, 194
Wallace, John Herbert 117
Wallace, L.A. 7
Wallace, Mary Ann xxv, 10, 13, 17, 19, 21, 24–5, 41, 50, 51, 55, 62, 81, 118, 156, 179, 189, 256
    residence at 44 Albany Street 10, 51
Wallace's giant bee 184
Wallace's golden birdwing butterfly 184
Wallace's standardwing 185
Wamar Island 121, 296
Wanumbai River 121
Warzbergen 116
wasps 16, 18, 40, 103, 114
Water Lily schooner 66, 81, 97
Waterhouse, George Robert xxvi, 37, 42, 43, 67, 73
Waterworth, William 255, 278
Webster (Mary's father) 62
Weraff brig 28
Westwood, Jojn Obadiah 189
White, Adam 82, 211
Williams & Norgate booksellers 12
Wilson, Algernon 22, 69
Wilson, Thomas 22
Wokam 296
Wollaston, Thomas Vernon 162, 164, 225
Woodbury & Page photographers 244
Woodbury, Walter Bentley 244
Woodford, Miss 57

Xylophaga 137

Zoological Society of London 255, 279, 280, 283, 285